Urban Agriculture

The *Urban Agriculture* Book Series at Springer is for researchers, professionals, policy-makers and practitioners working on agriculture in and near urban areas. Urban agriculture (UA) can serve as a multifunctional resource for resilient food systems and socio-culturally, economically and ecologically sustainable cities.

For the Book Series Editors, the main objective of this series is to mobilize and enhance capacities to share UA experiences and research results, compare methodologies and tools, identify technological obstacles, and adapt solutions. By diffusing this knowledge, the aim is to contribute to building the capacity of policy-makers, professionals and practitioners in governments, international agencies, civil society, the private sector as well as academia, to effectively incorporate UA in their field of interests. It is also to constitute a global research community to debate the lessons from UA initiatives, to compare approaches, and to supply tools for aiding in the conception and evaluation of various strategies of UA development.

The concerned scientific field of this series is large because UA combines agricultural issues with those related to city management and development. Thus this interdisciplinary Book Series brings together environmental sciences, agronomy, urban and regional planning, architecture, landscape design, economics, social sciences, soil sciences, public health and nutrition, recognizing UA's contribution to meeting society's basic needs, feeding people, structuring the cities while shaping their development. All these scientific fields are of interest for this Book Series. Books in this Series will analyze UA research and actions; program implementation, urban policies, technological innovations, social and economic development, management of resources (soil/land, water, wastes. . .) for or by urban agriculture, are all pertinent here.

This Book Series includes a mix of edited, coauthored, and single-authored books. These books could be based on research programs, conference papers, or other collective efforts, as well as completed theses or entirely new manuscripts.

Dona Pickard

Editor

Urban Agriculture for Improving the Quality of Life

Examples from Bulgaria

 Springer

Editor
Dona Pickard
Institute of Philosophy and Sociology
Bulgarian Academy of Sciences
Sofia, Bulgaria

ISSN 2197-1730   ISSN 2197-1749   (electronic)
Urban Agriculture
ISBN 978-3-030-94742-2   ISBN 978-3-030-94743-9   (eBook)
https://doi.org/10.1007/978-3-030-94743-9

This Springer imprint is published by the registered company Springer Nature Switzerland AG
The registered company address is: Gewerbestrasse 11, 6330 Cham, Switzerland

Foreword

Given the global challenges of urbanization, limited resources and food security, within the last two decades, urban agriculture (UA) has turned from being a peripheral discipline to a globally recognized mechanism for sustainable development. Within its "Growing Greener Cities" initiative, the Food and Agriculture Organization of the United Nations (FAO) has addressed UA as a recognized urban land use and economic activity, integrated into national and local agricultural development strategies, food and nutrition programmes, and urban planning. In Europe, a milestone in conceptualizing UA has been the COST Action Urban Agriculture Europe (Lohrberg et al., 2016). This EU-funded action networked practitioners and scholars interested in UA all over Europe. Its work is meanwhile taken forward by another EU Horizon 2020 project called "European Forum on Urban Agriculture, EFUA" and also by many local research and innovation activities, e.g. in Oslo, Madrid or the Ruhr area. However, activities in cities concentrate mostly on Western European countries, while in Eastern Europe, there are only few examples of ambitious agendas in UA or even established UA policies and practices (Lohrberg, 2019).

All the more so – and as former chair of the COST Action UAE – I am delighted that in Sofia, the capital of Bulgaria, a series of studies on UA have taken place which are presented here in this volume. The Sofia studies are of special interest and value as they give manifold insights in an Eastern European city's experience, which has clearly been insufficiently studied so far. I am very grateful to all the authors for their pioneering work in applying the COST Action's multiscale and multifunctional UA thinking to a city which has not been in the focus of UA research yet. Their findings are an indispensable contribution towards a better comprehensive understanding of UA in Europe and also to achieving better policies in this field.

I am thankful to the authors for sticking to the COST Actions general line of thinking which is, regarding UA, not just as "the production of food within cities" but as a tool to achieve goals in other arenas than food production as well, such as health and well-being, and recreation, and also social issues like inclusion, cohesion and empowerment. Also, the economic opportunities of all local food businesses and the ecological benefits of UA are insufficiently examined. It is a convincing and promising approach, introduced by Mariana Draganova and Dona Pickard, to address UA as a tool to enhance the "quality of life (QoL)" in our cities. Of course, enhancing the QoL is a goal no one would cast doubt on. However, this book is able to systematically link this umbrella approach to UA by offering a targeted set of methods for mapping, monitoring and better planning. Obviously, the QoL approach fits the widespread qualities of UA well, and especially its potential to unfold individual talents of self-development and community building.

To this end, it is quite useful to take a multi-disciplinary perspective as is done here. The authors reflect upon UA from four different disciplinary analytical points of view – social, economic, environmental and spatial – thereby following the general outline of the COST Action UAE. In addition, it is a particular asset of this book that the authors address various levels of UA. Too often, studies focus on the level of individual gardeners, but this book's chapters also elaborate on households and farms and, whenever relevant, also on neighbourhood or city level. This multi-level approach – also highlighted by the new Leipzig Charter on Sustainable European Cities (Council of EU ministers, 2020) – has proved its ability to deliver tailor-made policy recommendations, as in this volume.

This book is characterized by a high standard in combining qualitative and quantitative research methods. The former allow reflection on the multitude of UA representations in urban life and also to aggregate findings towards intricate factors for failure or success. The latter provide some robust data on links between UA and specific quality-of-life indicators which in the future will help to monitor the development of UA and adjust policies accordingly. It is really an achievement of this book to present a literature-based short list of quality of life indicators most relevant to UA.

The self-assessment tool for individual practices, adapted from the COST Action UAE, is also an asset to practitioners and municipalities who are willing to monitor how UA initiatives contribute to community or city quality of life. The initiatives themselves will certainly benefit from this tool to better adjust their agendas. As in this self-assessment tool, the whole book gives evidence that UA can only be managed towards a better quality of life with the genuine participation of and dialogue between all stakeholders. It is clearly shown that those citizen groups that could benefit most from UA – like disadvantaged and vulnerable groups, or children and children's institutions – are often the ones critically needed to engage in UA governance. To overcome this deficiency, the volume outlines specific discrepancies between civil participatory methods to develop UA for improved quality of life and top-down policy patterns that hinder this development.

While the presented case outlines the present situation in Sofia, the gathered knowledge and data is also highly relevant to other locations, especially in Eastern Europe with its multifaceted post-socialist development of cities. UA in these cities is still often neglected by policymakers. However, UA can adhere to specific roots different from other European regions, such as the tradition of family gardens and a strong rural-based knowledge in family-based, small-scale agriculture which many migrants brought to the growing and industrializing cities – and which was also often tolerated or even supported by former socialist regimes, as shown in the case of Sofia.

Hence, it is now the time to restore and reinvent the manifold links between cities and citizens and UA – also and foremost in Eastern European cities. To conclude, let me express my hope – and my confidence – that this volume will serve as a milestone in promoting UA in Sofia and also aid in making better UA policies all over Europe. By revealing success factors and barriers for the development of UA, this book will inform and inspire a broad range of UA stakeholders – practitioners and researchers, policymakers and activists, farmers and grassroots participants – to sustain their UA activities and thereby improve the quality of life in our cities.

Head of the Institute for Landscape   Frank Lohrberg
Architecture, RWTH Aachen
University, Aachen, Germany

References

Lohrberg, F., Licka, L., Scazzosi, L., & Timpe, A. (Eds.). (2016). *Urban agriculture Europe*. Jovis.

Lohrberg, F. (2019). Urban agriculture forms in Europe. In E. Gottero (Ed.), *Agrourbanism. Tools for governance and planning of agrarian landscape* (pp. 133–147). Springer.

Council of EU ministers. (2020). *The New Leipzig Charter. The transformative power of cities for the common good*. Retrieved October 15, 2021, from https://ec.europa.eu/regional_policy/sources/docgener/brochure/new_leipzig_charter/new_leipzig_charter_en.pdf

Acknowledgements

First and foremost, we would like to acknowledge the financial support which we received from the Bulgarian National Science fund, without which we would not have been able to carry out our interdisciplinary research project "Urban agriculture as a strategy of improving the quality of life of urban communities", the results of which served as a basis of this book.

We also want to extend our gratitude to all urban gardeners, civil organization activists, education staff, municipality officials, urban farmers and research partners who gave their time to welcome us in their gardens, fields and offices, answer our questions, and participate in focus groups, discussions, an online conference and a round table. Their active participation in the project allowed us to gather diverse and rich data that inspired us and confirmed our belief that urban agriculture in Sofia has a fruitful ground to develop on, in a way that serves urban communities and the environment.

Special acknowledgement is extended to the municipal enterprise SofiaPlan who graciously gave us permission to use data and materials from the Urban Masterplan for Sofia Municipality – maps and boundaries of the different development zones, main transportation arteries, and parks and other elements of the urban environment.

Throughout the project execution, we relied on the administrative and technical support of the Institute of Philosophy and Sociology, which we highly appreciate.

Contents

Abbreviations

BG at SU	Botanical Garden at the Sofia University
EEA	Environment Executive Agency
GDP	General Development Plan of the City of Sofia
MAC	Maximum allowable concentration of harmful substances in soil
MoEW	Ministry of Environment and Water
NGO	Non-governmental organization
PC	Precautionary concentration of harmful substances in soil
QoL	Quality of life
TEFS at UF	Training and experimental field station at the University of Forestry
UA	Urban agriculture
UDC	Utilitarian-decorative coefficient of urban agriculture

Chapter 1
Introduction

Mariana Draganova and Dona Pickard

Abstract In the introduction we present the topic of urban agriculture (UA), as well as the relevance and rationale of the research into the effects of UA on the social, economic, environmental and spatial aspects of the quality of life (QoL) in Sofia, Bulgaria from the study Urban Agriculture as a Strategy for Improving the Quality of Life of Urban Communities, conducted in Sofia between 2018 and 2021. The analyses are based on qualitative and quantitative studies (in-depth interviews, focus groups, expert statements, and a sociological survey representative of three districts of Sofia among 750 people), as well as on ecological monitoring of UA sites and spatial mapping of natural UA resources in the City of Sofia. In the introduction we also briefly present the structure of the book.

1.1 About the Topic

Urban agriculture (UA) is generally viewed as an agricultural activity of citizens and urban communities that helps improve the quality of their life and contributes to sustainable urban development by alleviating a range of pressing problems of modern urban life, such as increasing urban poverty, lack of social cohesion, food and energy insecurity and greater pressure on the environment in cities. Worldwide, agriculture in the cities has been accepted and developed as a socially significant activity that has positive effects on communities and on the environment (Vejre & Simon-Rojo, 2016). When it is integrated in the urban environment and eco-system, it can contribute to the sustainability of cities (for example by using organic waste as compost, stimulating short food supply chains, creating and maintaining green infrastructure, etc.). It can also improve the quality of life (QoL) in cities by providing business opportunities for groups at risk of poverty, creating opportunities for active recreation and alleviating the effects of climate change in cities. In these ways, UA can be integrated in the economic, social and ecological urban systems and, in many locations around the world, is becoming an inseparable part of urban

M. Draganova · D. Pickard (✉)
Institute of Philosophy and Sociology, Bulgarian Academy of Sciences, Sofia, Bulgaria

© The Author(s), under exclusive license to Springer Nature Switzerland AG 2022   1
D. Pickard (ed.), *Urban Agriculture for Improving the Quality of Life*, Urban
Agriculture, https://doi.org/10.1007/978-3-030-94743-9_1

life and a carrier of new trends in the life of urban communities. Urban agricultural practices abound around the world: both in cities marked by financial and economic crises, as well as in more economically stable cities. It encompasses and creates linkages and relationships between local economic actors, civil society and local authorities, and contributes to the development of sustainable local food alternatives and communities with greater solidarity and integration.

In the European context the issue of UA is also relatively well developed, both as various practices and as a research focus. In some countries which have faced serious economic crises since 2008, for example Greece, there has been an increase in the number of communal vegetable gardens to provide food security for people at risk of poverty (with low or no income, pensioners, unemployed, single parents, etc.) and UA is already a part of municipal social policies that give a new direction to urban planning (Partalidou & Anthopoulou, 2017). In Holland, UA is often implemented to strengthen the social cohesion in neighbourhood communities, involving the citizens in commonly grown food and greening the environment (Veen & Eiter, 2018). In London the local authorities recognize UA initiatives as an important contribution to making citizens feel less isolated, safer, mentally and physically healthier, as well as having many environmental benefits, including enhancing the green infrastructure and providing habitats for the city's biodiversity (Greater London Authority, 2018).

Since the 1990s, the problems of the environment and food security in the realm of the social sciences have become very significant, not only because of the ubiquitous effects of global climate change on everyone's personal life, but also because of the richness of important social phenomena and processes that the relationship between humans and nature reflect. These include the number of problems of the modern risk society such as poverty, inequalities and social exclusion; distribution and access to natural and power resources; the link between responsible consumption and local and global transformations in the models of production and consumption of food; collective action and trust, solidarity and social justice, civil activity and participation.

Since 1972, when the earliest article on UA was published (ETC, 2001), the focus of most studies on the topic has been on one main aspect of UA and commonly that has been done on the basis of case studies of individual practices. The need for reliable and in-depth fundamental research on the potential of UA for improving the quality of life of urban communities that incorporates an analysis of all its functions and effects on society and nature was recognized and reflected in the funding of a European multinational and multidisciplinary research action under the European commission COST programme – Urban Agriculture Europe, which ran from 2012 to 2016. The results of the action included a set of reports on the social, economic, environmental and spatial effect of UA activities which are seen universally across the European cases that were studied within the action. Moreover, the researchers who were involved in the action came up with a joint declaration on the importance of informing urban policies with scientific research in the field of UA (Lohrberg et al., 2016).

This declaration, together with the fact that no scientific research had been done in Bulgaria and in particular in Sofia, on the existing forms of UA and their effect on urban life since the end of the critical first ten years of transition from a socialist to a democratic regime, inspired the multidisciplinary team of authors of this volume to undertake a comprehensive study of the practices of UA in Sofia, the capital city of Bulgaria, and to see if these activities have the potential to truly improve the QoL of urban communities as demonstrated in other studies from around the world. Our idea was supported with research funding from the Bulgarian National Science Fund[1] and implemented within the frame of a three-year research project that ran from 2018 to 2021, named "Urban Agriculture as a Strategy for Improving the Quality of Life of Urban Communities". The results of the project revealed success factors and barriers for the development of UA as a tool for enhancing QoL that are linked to social and economic system factors and environmental and spatial contexts that could be relevant and enlightening for other locations beyond Sofia and Bulgaria. Thus, we prepared this volume to present our study and its outcomes to a broader audience of practitioners, researchers, as well as policy and civil organizations that might find the links we make between different aspects of UA to the various dimensions of QoL relevant to their studies or work.

1.2 Relevance and Aims of the Study

In Bulgaria urban agriculture (UA) is slowly gaining popularity among actors who were not traditionally involved in agricultural activities, but in the public discourse UA is still exploited only as an extravagant idea, seen in occasional initiatives of non-governmental organizations, civil groups and individuals creating vegetable patches in kindergartens, urban yards and other public spaces (Koleva & Pickard, 2015). These practices do not have any official recognition in institutional visions, strategies and policies on urban planning, education, ecology and social cohesion. This is why they are left outside of the control and regulations of the institutions. The significant contribution they could have for improving the socio-economic status of urban communities is therefore left unexplored.

Urban agriculture is not developed as a research field in Bulgaria either, and it has not been a central subject of historical, ethnological, sociological, environmental or urban studies. This motivated us to initiate and conduct a study based on a holistic and multidisciplinary approach to the subject. It explores the forms and practices of UA with a potential for having a social, educational or environmental impact on the quality of life of the citizens of Sofia. In Bulgaria agrarian activities are still associated with rurality and backwardness, so our project is a challenge for the public, institutions and local authorities to rethink this stereotype, to broaden the horizons and perceptions of UA practices by gathering scientific evidence that can help establish the specific mechanisms and tools to be used to maximize the benefit UA can have for a more sustainable development of the urban environment in Sofia and a better life for its communities.

---

[1] Under Grant agreement DN05/18 from 17 December 2016.

The research methodology is based on a variety of methods – socio-economic (qualitative and quantitative) methods, spatial-territorial surveys and ecological evaluation. These are presented in greater detail in the following chapters.

1.3   Structure of the Book

Chapters 2 and 3 frame the case-study context and the conceptual foundations of the research and are presented in Part I of the book.

To acquaint the reader with the current state of UA development in the case-study city of Sofia, Chap. 2 gives a short overview of the historical evolution and current characteristics and challenges that UA faces in Sofia and Bulgaria. It sketches the social and cultural specifics that shape the public attitudes towards UA in the city and the local authorities' approach to the UA phenomenon at present. It also briefly describes the geographic and climate conditions for UA development in Sofia.

The conceptual links between the dimensions of QoL and UA activities are presented in Chap. 3 as the first task in the study. In this chapter we discuss the relationships between those indicators of urban community QoL that are most commonly affected positively by UA activities according to the scientific literature. We set the boundaries of the research, and define the concepts and indicators we have worked with. While keeping in mind the distinction between opportunities and real outcomes of a good life, as well as the external conditions and the individual dimensions of QoL, we present the structure of the QoL indicators at the individual, local community and city level. This chapter also presents a short overview of the research methods used in our study: the qualitative (in-depth interviews, focus groups, expert statements) and quantitative studies (sociological survey representative of three districts of Sofia among 750 people); the ecological monitoring of UA sites; and the spatial mapping of natural UA resources in the City of Sofia.

Part II presents our study on the dimensions of UA impact on the QoL in Sofia in detail and is divided into five chapters following the design of the research project. This is based on four disciplinary approaches to the potential effects of UA on the QoL of urban communities.

Chapter 4 explores the social dimensions of UA impact on QoL in the capital city. It looks into the way social cohesion and inclusion could be enhanced through urban agriculture practices, the educational effects of UA, and how sustainable consumption patterns can be established with the help of UA.

Chapter 5 is dedicated to the economic impacts of urban agriculture at various levels: its potential to promote the development of local food business, as well as the economic effects of urban agriculture for individuals, households and society as a whole.

Chapter 6 presents data and conclusions on the potential UA has to improve the environmental conditions in urban areas and some of the factors that determine its success in achieving positive environmental outcomes.

Chapter 7 illustrates the spatial dimensions of QoL in the city which are directly affected by UA. This chapter presents a wealth of illustrative material, collected through field research and satellite data analysis from three districts of the city with very different landscape profiles.

The most important findings and conclusions of our research are summarized in Chap. 8.

From the lessons learned from each chapter in Part II, and the key indicators we have found to be determining the extent to which UA can help improve the QoL of urban households and communities, we have devised a self-assessment tool that can be used by practitioners, civil organizations and policy-makers when evaluating the importance of any given UA practice for the QoL of urban communities. This tool is presented in Appendix.

References

ETC Foundation. (2001). *Annotated bibliography on urban agriculture*. Retrieved June 6, 2021, from https://ruaf.org/assets/2019/11/Annotated-Bibliography-on-Urban-Agriculture.pdf

Greater London Authority. (2018). *The London food strategy. Healthy and sustainable food for London*. Retrieved June 6, 2021, from https://www.london.gov.uk/sites/default/files/final_london_food_strategy.pdf

Koleva, G., & Pickard, D. (2015). Sustainable development values in urban land use. In E. Marinova (Ed.), *Ecological ethics. Nature and sustainable development* (pp. 379–391). Sustainable Development for Bulgaria Foundation. (In Bulgarian).

Lohrberg, F., Lička, L., Scazzosi, L., & Timpe, A. (Eds.). (2016). *Urban Agriculture Europe*. Jovis.

Partalidou, M., & Anthopoulou, T. (2017). Urban allotment gardens during precarious times. From motives to lived experiences. *Sociologia Ruralis, 57*(2), 211–228. https://doi.org/10.1111/soru.12117

Veen, E. J., & Eiter, S. (2018). Vegetables and social relations in Norway and the Netherlands. A comparative analysis of urban allotment gardeners. *Nature and Culture, 13*(1), 1335–1160. https://doi.org/10.3167/nc.2018.130107

Vejre, H., & Simon-Rojo, M. (2016). Phenomenon: Introduction. In F. Lohrberg & L. L. A. Timpe (Eds.), *Urban agriculture Europe* (pp. 16–17). Jovis.

Part I
Overview of the Sofia Urban Agriculture Case and Research Approach

Chapter 2
Historical Evolution of Urban Agriculture Practices in Sofia and Local Context at Present

Mariana Draganova and Dona Pickard

Abstract Since 1879, when Sofia was declared the capital of Bulgaria – just liberated from Ottoman rule – urban agriculture in the city has gone through various modes of existence and functions, but has consistently been a part of city life for individuals and households. Families in the city have relied on producing food for self-sustenance during crises, which have marked the unsteady economic development and changes in the political regime. The individual and private orientation of UA in Sofia has not changed significantly and there are few UA practices addressed to solving broader urban problems at the community or city level. These practices are currently carried out by a small number of individuals and organizations, without active regulatory support from Sofia Municipality.

2.1 Historical Background

The historical development of Sofia as Bulgaria's capital and as a social organism was not a smooth process and has been marked by a series of political turns that sharply changed the social and economic profile of the city at intervals of about half a century apart. Since 1879, when Sofia was declared the capital of Bulgaria – just liberated from Ottoman rule – UA in the city has gone through various modes of existence and functions.[1] The population of the newly established capital nearly doubled in its first year of being the political and administrative centre of the country – from 11,000 to 20,000 inhabitants, thanks to the newly arrived population. In another 20 years, in 1900, the population was already 68,000 and in 1946, at the beginning of the socialist regime, it was 435,000. Although in the few first decades after it became the capital, Sofia saw enormous changes in urban planning and a striking architectural makeover, the agricultural habits of its citizens and the rural

---

[1] For the period before the Liberation, see Chap. 7.

M. Draganova · D. Pickard (✉)
Institute of Philosophy and Sociology, Bulgarian Academy of Sciences, Sofia, Bulgaria

© The Author(s), under exclusive license to Springer Nature Switzerland AG 2022
D. Pickard (ed.), *Urban Agriculture for Improving the Quality of Life*, Urban Agriculture, https://doi.org/10.1007/978-3-030-94743-9_2

traditions that the newcomers also brought to the new capital, kept urban agriculture as an inextricable part of Sofia's daily life.

Like other Bulgarian cities at that time, Sofia's economy, spatial planning and demography did not differ dramatically from rural settlements, although crafts, commerce, and modern-day industry were the basis of the urban economy. Historical documents and essays testify to the abundance of agricultural land and UA practices in urban areas at the end of the nineteenth century, which may be called archetypes of modern urban agriculture. These included areas within the built-up parts of the city which could be used for agricultural production and were fully intertwined in the urban environment. Among the most common agricultural properties were the vegetable gardens and orchards of urban houses, followed by the "bostan" – plots planted with melons, watermelons and pumpkins, some of which were also in the built-up areas of the city (Figs. 2.1 and 2.2).

Built-up areas were surrounded by fields, meadows, vineyards and more vegetable gardens (Draganova et al., 2018). Livestock, especially small livestock such as poultry, was popular (see Fig. 2.3). Even people working at the newly established House of Parliament kept chickens in a parliamentary outbuilding in the first years of their service (Tahov, 2015). In the 1930s the urbanized territory of Sofia was expanding fast towards neighbouring villages. With the establishment of the Great Urban Municipality (1938), nine villages were officially joined to Sofia. These peripheral rural areas, which incorporated elements of urban lifestyle and architecture, still kept their agricultural profile, given the presence of agricultural land around the houses of the local population.

The migration of a rural population to large urbanized centres was not only a mechanical process of mobility but also a profound socio-psychological breakthrough in their consciousness and behaviour, a long process of adaptation of their rural values, lifestyle and culture to the urban environment and way of life. The new urban dwellers of rural origin were looking for plots of land for construction, but also for land that could be a source for food production, which could help preserve the traditions of growing food for the household (Draganova et al., 2018). This is why urban centres, albeit following the path of modernization, kept agriculture as a thriving part of their existence, sustained by the local households.

In the early twentieth century agriculture was slowly moving along the capitalist path of development. Land was privately owned, concentrated mainly in small and medium-sized farms. After the end of the Second World War, with the change of the political regime, the socialist government introduced a series of reforms in land ownership and land use, as well as in agriculture and the economy, bringing about another fundamental change in agriculture and urban life. Collectivization of the land began in 1947 and large-scale collective farms were formed. In the late 1950s they constituted more than 90% of the production units in the sector. Over the next 40 years of the centrally planned economy, agriculture was oriented to food supply and food security based on a centralized market, distribution, export and import.

Agriculture was increasingly managed in a top-down way, with continuous experimentation of different farming models. Since the beginning of the 1970s, the collective farms, which had functioned largely as cooperatives until then, were integrated into agro-industrial complexes in order to achieve a higher degree of

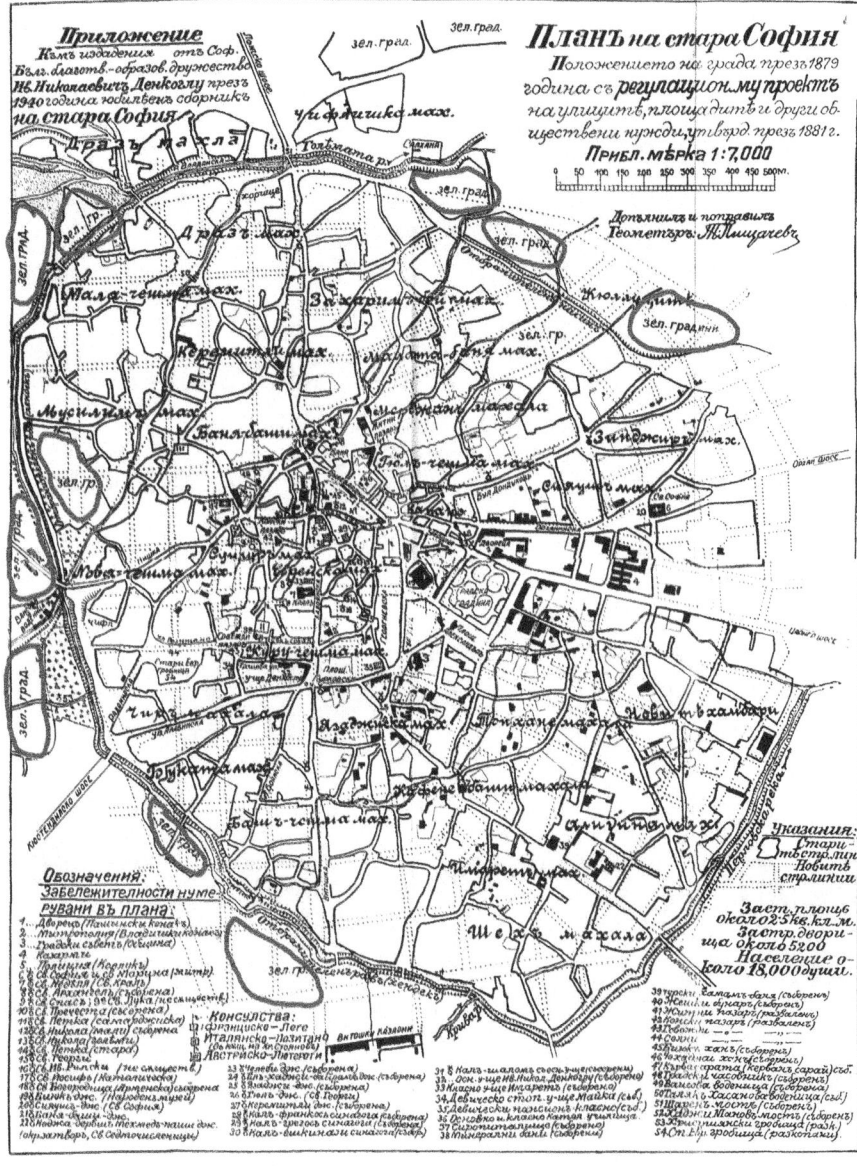

Fig. 2.1 A map of Sofia from 1881. The locations of the vegetable gardens are encircled in red

concentration of production and specialization. However, these changes resulted in a persistent reduction in production, weakening the efficiency of the agro-food chain links and a very serious social impact on the rural population, which became demotivated to work in the sector and started a large migratory wave from villages to cities, including Sofia (Draganova et al., 2018, Pickard, 2013).

Fig. 2.2 A brewery and garden near Sofia centre, c. 1900

Fig. 2.3 A Sofia household with their pigs from WWI

In the 1980s, the government introduced additional incentives to increase the profitability of the sector. Apart from partial decentralization of the farm structures and fiscal measures, the state introduced measures to provide land for growing fruit and vegetables and for animal breeding for self-subsistence. One of these measures was a series of state decrees leasing state or municipal land to families for "eternal use", as long as they cultivated it according to the rules (Pickard, forthcoming). Although the state claimed Bulgarian society had achieved "advanced socialism", which implies stronger collective forms of production, the private use of land for household food consumption, including in and around cities, was never abolished and remained very prominent both in the 1980s and in the 1990s when, after another sharp political turn – the fall of socialism – Bulgaria went through a great economic crisis. Data from the 1980s show that approximately half of Sofia households were engaged in cultivating fruit, vegetables and herbs and more than 90% made preserves and pickles from home-grown or purchased agricultural produce. The percentage of households in Sofia achieving self-sufficiency in various foods in that period (either self-produced or processed by friends and relatives in the country) was about 14% (Yoveva et al., 2000: 502). At the beginning and in the middle of the 1990s, when Sofia went through a food crisis, UA significantly contributed to the food security of citizens. UA did not have a significant market dimension during this period, as farming households produced almost entirely for self-consumption. While statistics show that 28% of the households earned some income from UA during these periods, this was mostly in-kind income (Nugent, 2000, Yoveva et al., 2000).

With the change of regime, land use changed considerably. Agricultural cooperatives and state farms were liquidated and land was restituted to its pre-socialist owners and, in most cases in the city, sold to new private owners for construction. The peri-urban villages south of the City of Sofia lost their agricultural character and became densely populated over the past two decades and large territories of fertile agricultural land were converted to housing estates (Draganova et al., 2018). Statistical data show that the share of agricultural land has been reduced by 5% over the decade between 1998 and 2009 (from 41% to 36% for agricultural land), and the share of urbanized areas has increased by slightly less than 3% (from 18% to 22%).[2]

Before Bulgaria's accession to the European Union in 2007, small private farms abounded on the borders of the city and in the neighbouring villages. They used private yards or privatized former state and collective farms and produced vegetables, fruit and animal produce both for their own consumption and for the market. These farms were on a very small scale, usually not having more than two or three cows, five goats and a few other small animals like pigs, etc. (Yoveva et al., 2000: 505).

With the introduction of stricter rules for production, processing and selling food produce, related to the new status of an EU member state, Bulgaria, including the capital city, lost most of these semi-legal production units. Stricter regulations on the movements of edible products were introduced, as well as new and stricter controls on the traders and markets. The insecure statute of land ownership and unstable newly established land market further hindered the modernization and sustainable development of UA in Sofia. Regulatory institutions and the local government, although having available highly qualified staff for urban planning, monitoring land use, construction and environmental protection, as well as veterinary and

---

[2] Authors' own calculations based on data from Recasens et al. 2016 and Yoveva et al. 2000.

sanitary food control (ibid, 512), did not take any specific measures to encourage UA practices to adapt to the new conditions and to contribute to the city's food security. The predominant forms of urban and peri-urban agriculture remained gardening and small livestock breeding for self-sustenance and sometimes informal trade in back-yards and private gardens adjacent to family houses, in the gardens of second homes or summerhouses, usually not further than 100 km from the city centre and on plots of land distributed to citizens with the aim of providing self-supply of food under state decrees issued from the early 1970s until the mid-1980s (ibid. 506).

As early as the end of the 1990s, researchers pointed to the various potentials of Sofia in developing UA practices that support the social, environmental and eco-nomic development of the city, including the use of neglected urban terrains, the strong linkages between urban households and the countryside, and the preserved knowledge of agricultural production on a limited scale. Although it was more than 20 years ago that scientists underlined the need for municipal strategies to "further promote private initiative in urban farming", including cheaper credits for UA, creation of a municipal agricultural investment fund and reduction of the production risks (Yoveva et al., 2000: 514), no progress has been made in the direction of establishing a comprehensive city strategy for UA or for the city food system.

2.2   Local Context at Present

While UA has had a long history and traditions in Sofia and other Bulgarian cities, and even today a significant share of citizens in the capital have a strong connection to agrarian activities, whether by gardening themselves or helping relatives in the countryside who provide them with part of their food throughout the year, UA is absent from local and national policies. UA is not recognized as a tool either for the food security of the city, neither is it seen as a mechanism to improve any other aspect of urban life. It is seen by the policy-makers as a private activity beyond the regulative responsibilities and power of the municipality. The few civil initiatives set up to develop UA as a collective and socially significant activity in the city have not been able to convince the municipality authorities otherwise. Indeed, as our survey has shown, about a quarter of all respondents practice some form of UA, but of them, less than 5% do it for social reasons – to feel part of the community (See Fig. 4.1, Chap. 4).

At present, in Sofia there is still a good potential for developing UA that supports sustainable urban development and contributes to the quality of life of urban communities. While developing and growing as a contemporary urban centre, the city still retains urban and suburban areas with land and water that favour agricul-tural development. Sofia Municipality, with just over 1.3 million people in 2014, having grown by about 2% for 3 years (Sofia Municipality (b), 2000), is located at about 550 m above sea level, on a plain surrounded by mountains. The settlements and urbanized territories of the municipality occupy a little less than 250 km², out of the total 1311 km² of the area of the municipality, while the agricultural territories

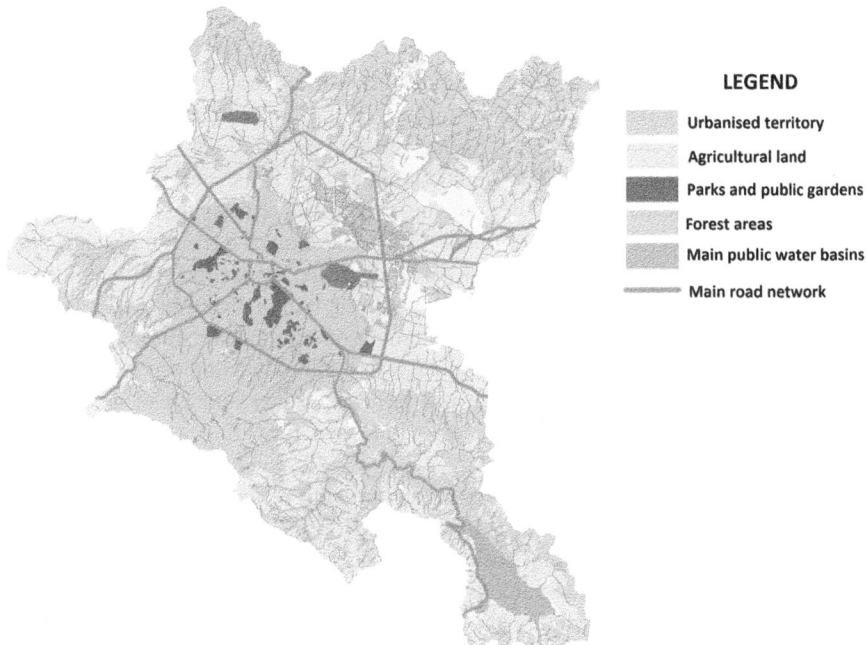

Fig. 2.4 Map of Sofia's agricultural land, parks and public gardens, forest areas and public water basins. (Obtained from GIS-Sofia)

around the city amount to 509 km². The rest of the territory is accounted for by forests, mining areas, transport and infrastructure and water basins, including eight rivers, the biggest of which, the Iskar, serves as the main drainage artery of the Sofia plain (Recasens et al., 2016, Sofia Municipality (a), 2020). A lot of wild green areas are scattered around Sofia and along the rivers that run through the city (Fig. 2.1) – these areas are called "green wedges" as they seem to cut into the city from its periphery to its centre (Fig. 2.4).

Major soil types of the Sofia plain are maroon sandy soils, alluvial and illuvial sand and clay and black earth (humus) (Yoveva et al., 2000: 501). The climate of Sofia, cooler than the rest of the region, allows for the production of more cold-tolerating crops such as cabbage, carrots, leeks, potatoes, apples. Bees are also very common in the area of the Sofia plain.

Sofia Municipality includes the City of Sofia, where most of the population resides, three towns and 34 smaller villages. Today there are no urgent food needs at a mass level for the citizens of Sofia whose requirements and expectations of the quality of food products are growing. The consumer needs, preferences and patterns towards cleaner and healthier food of trusted origin, produced under environmentally friendly conditions, are increasing and becoming more common. Thus, urban agriculture itself is slowly starting to transform from just being a "food bank", to a phenomenon with social, ecological and educational functions (Draganova et al., 2018).

Present-day Sofia is a modern, growing city with characteristic features of a heavily urbanized environment but with many green areas. As elsewhere around the world, a serious challenge facing Sofia's green system – along with a constantly increasing population (both permanent and temporary) – is the ever increasing need for areas for construction. Building in the peri-urban region, together with the physical destruction of the soil, results in a fragmentation of agricultural land, thus imposing limitations on the working of the land. At the same time, as the evidence from the interviews and expert opinions gathered within the framework of our project shows, the capital city has quite a number of unattractive areas between the residential blocks, as well as abandoned, neglected municipal and private plots that are not being looked after, but which can be transformed into green areas, both implementing the practices of urban agriculture and providing a source of food (ibid.).

While from a European perspective, urban agriculture can be seen not only as a source of fresh food but also as a mechanism for social integration, economic development, and environmental sustainability implemented in many forms such as community and educational gardens, vertical and rooftop farms, in Bulgaria and in Sofia such practices are still rare and the traditional forms of UA dominate. These include:

- **micro-farming for personal/family use** near the house of residence (family gardens). They are usually small plots of land in the backyards of households or near the city where people produce vegetables and fruit. Almost invariably the food produced in this way is not enough to provide for all the nutritional needs of a family, so many people regard this activity purely as a leisure and recreational farming activity to combat stress, as it does not provide any monetary income. For many city-dwellers these urban agriculture practices are perceived as such an integral part of life that it is not regarded as "agriculture", since it is not an entrepreneurial undertaking. People share the opinion that they like to feel connected to nature, in addition to enjoying the personal satisfaction of having produced fresh, clean and healthy food themselves.
- **local food farms and entrepreneurial models of UA** in compliance with the EU regulations have started appearing on the periphery of the municipality to utilize their proximity to the city and direct their production to Sofia's urban market. Their aim is to address the consumer demand for the higher price segment of ecological and organic produce. In recent years, the niche markets and consumer expectations for local, seasonal food produced in a sustainable way have been evolving, which is a push factor for peri-urban farm development and local food supply. These are mainly small family farms that create and maintain their own consumer network which they supply directly and on a regular basis, through a mailing list including consumer cooperatives and markets. Several farmers' markets have been established in Sofia with the local municipality's blessing but not in a legal normative frame because the concepts of farmers' markets and

UA are still not integrated in the current normative framework for urban development or urban planning.

- newer and rare models of urban agriculture practices in Sofia are based on **collective actions** such as community and educational gardens. Participants consider joint gardening activities as helpful in building communities by establishing personal contacts, deepening the feeling of trust between individuals, creating close personal connections and an atmosphere of solidarity that lead to social action in other spheres outside UA. Five **community grassroots gardens** have been initiated on abandoned and neglected land in various Sofia areas since 2015, two of them have stopped functioning. In the rest of them people meet and cultivate vegetables and herbs in a collaborative way, learning together and sharing their experience, knowledge and tools. Because of lack of any official policy support, engaged local communities face permanent uncertainty about whether they will be removed from the plot. For them, it is important to share what they produce, but most important of all is the connection with the land, social networking and the feeling that they are in charge of the public space. Community garden development is inspired by new, post-material concerns about health, quality of food and the environment. There are also new actors in these developments – young, educated and well-off citizens who are boosting the demand for clean food and responsible consumption models and lifestyles.
- **small educational gardens** based on NGO projects have been initiated at some kindergartens and schoolyards in Sofia. These projects focus on opportunities for children to have daily, fulfilling and educational contact with live nature when they are personally involved in vegetable gardening at their kindergarten or school. There, children can learn, work, observe, create and study the principles of nature and how to cooperate with others.

In Sofia, UA faces a number of challenges, such as pressure on open space and farmland, barriers to cooperation with more traditional farmers, lack of entrepreneurial skills, achieving and maintaining profitability, lack of finance and lack of supportive policies, and sources of pollution arising from industrial activity. What drives the socially-significant development of UA at present is a mix of civil initiatives that work in parallel and enforce each other. One of them is the civil "Initiative for the Development of Urban Agriculture in Sofia", uniting urban gardeners, citizens who are looking for community gardens in the city, researchers and NGO activists, all aiming to put civil pressure on the municipality to recognize UA as a legitimate urban activity with various benefits to city life, to regulate its basic forms and ensure terrains for it. Local authorities have not taken any official measures to encourage UA in Sofia, apart from voting for a 30-year long-term strategy for urban development called "Vision for Sofia", which includes measures to protect high quality agricultural soils and water sources and support UA practices that contribute to the QoL of urban dwellers. A very fruitful partnership was established between the civil initiative for the development of UA in Sofia and the technical team of the municipality enterprise SofProject, which researched and developed "Vision for Sofia" – they collaborated actively to create an online interactive map that identifies 1845 urban terrains across the city that are suitable for UA. Although identifying a spot on the map does not mean the owners (private or public) will agree to it being used for UA, this collaboration is one of the few partnerships that showed civil groups can work with municipality representatives to produce results that are beneficial for all the stakeholders.

This lack of sustained dialogue between all stakeholders, particularly between civil groups and the municipality, reflect a fundamental problem with social cohesion and genuine civil participation in the governance of the capital and the decision-making processes that ultimately affect all citizens. In Sofia, while the average income is higher than in the other parts of the country, inequalities are also deeper and there are internal regional inequalities and concentration of poverty and social exclusion in certain districts of the city (SofProject, 2019). Thus, while there is a host of socio-economic problems in the city that would best be resolved through a continuous and constructive dialogue between representatives of all social actors, there is an almost impermeable barrier between a non-citizen oriented municipality administration and policy and the rest of the civil actors. This situation does not seem to have an obvious solution at present as the general levels of trust in Bulgarian society are very low, both in terms of interpersonal trust and trust in institutions (Eurostat, 2021), and it is difficult to start and maintain dialogue between different actors, especially where they stand at different levels of power.

Although at the moment the municipality is governed with little genuine citizen participation (Pickard, 2020), the pressure civil society is putting on the decision-making process cannot be denied and it could be expected that slowly the local institutions will start considering the arguments of civic groups, and engage in constructive dialogue with them. The civil pressure groups that have demands on the municipality are not numerous, but they are well organized and rather persistent in their attempts to engage the municipality in a dialogue about UA. Their demands include both long-term policy-based regulations that recognize UA as an activity that is important for the city and is supported by the municipality, but also infrastructure and material support, including the provision of terrains for urban gardens, financial support for purchasing water containers for rain water, sheds and tools, etc., especially for educational and social institutions. Additionally, UA activists expect the municipality to fund research into how suitable different municipality terrains are for UA, by testing soils and water sources. Local authorities are also expected to organize or actively support awareness campaigns among the citizens and to inform them about the benefits of UA. Still, the most fundamental demand that active UA supporters have on the municipality, is that the public interest is just as important as the financial dimensions of any municipal decision on how public land and other resources are used. At the moment the shared perception of the civil sector is that local governance is based on the most profitable in monetary terms and not on what is best for the common good of the citizens of the capital

Overall, the biggest problem of establishing effective partnerships between the local authorities and the UA practitioners and activists, is the lack of trust in the local institutions and the perception that the interests of all stakeholders are not represented fairly when the municipality realizes them with regard to UA policies, but that they are created and applied as a result of unlawful dependencies and that local authorities serve the interests of a narrow circle of financially strong organizations (Pickard, 2020).

To sum up, while the main motivation to engage in UA in the past was connected to economic needs (traditional livelihood, self-subsistence, income substitution), the present revitalization of UA practices in Sofia is based on social and environmental goals. Understanding the change in the motivation to initiate and participate in UA activities from the past to the present is crucial if the potential benefits of these activities are to be explored to the fullest extent. This understanding, together with an analysis of the barriers to UA development, can provide the answers to how to stimulate UA practices that are more socially, economically and environmentally beneficial for the city.

References

Draganova, M., Koleva, G., & Pickard, D. (2018) *Urban agriculture in Bulgaria: Legacies and current development*. Report presented at 14th international conference on urban history, August 29th–September 1st, Rome, 2018.

Eurostat. (2021). *Average rating of trust by domain, sex, age and educational attainment level*. Retrieved June 12, 2021, from http://appsso.eurostat.ec.europa.eu/nui/show.do?dataset=ilc_pw03&lang=en

Nugent, R. (2000). The impact of urban agriculture on the household and local economies. In M. Bakker, S. Gündel, U. Sabel-Koschella, & H. de Zeeuw (Eds.), *Growing cities, growing food: Urban agriculture on the policy agenda* (pp. 67–98). German Foundation for International Development.

Pickard, D. (2013). *Social capital as a factor for development of rural communities in Bulgaria*. PhD dissertation, "Prof. Marin Drinov" Academic Publishing House.

Pickard, D. (2020). The attitudes of local authorities and civil actors to Sofia's green urban development through urban agriculture. *Sociological problems, 52*(2), 575–595.

Pickard, D. (forthcoming). The role of centralised policy planning for Bulgarian urban agriculture heritage from the socialist period. In F. Lohrberg & A. Timpe (Eds.), *Urban agriculture heritage*. Birkhäuser.

Recasens, X., Sjöblom, J., Koleva, G., & Pickard, D. (2016). Sofia. A city with urban agriculture potential. In F. Lohrberg, L. Lička, L. Scazzosi, & A. Timpe (Eds.), *Urban agriculture Europe* (pp. 149–155). Jovis.

Sofia Municipality (a). (2020). *Sofia geographical information*. Retrieved June 12, 2021, from https://www.sofia.bg/web/tourism-in-sofia/geographic-characteristics-of-sofia

Sofia Municipality (b). (2000). *More information about Sofia*. Retrieved June 12, 2021, from https://www.sofia.bg/web/tourism-in-sofia/more-information-on-sofia

SofProject. (2019). *Vision for Sofia. Report on people*. Retrieved June 12, 2021, from https://vizia.sofia.bg/wp-content/uploads/2018/01/Доклад_Хора.pdf. (In Bulgarian).

Tahov, R. (2015). *The big parliamentary spectacles. A chronicle of the scandals in the national assembly*. Litus (In Bulgarian).

Yoveva, A., Gocheva, B., Voykova, G., Borissov, B., & Spassov, A. (2000). Sofia: Urban agriculture in an economy of transition. In M. Bakker, S. Gündel, U. Sabel-Koschella, & H. de Zeeuw (Eds.), *Growing cities, growing food: Urban agriculture on the policy agenda* (pp. 501–518). German Foundation for International Development.

Chapter 3
Framing the Research Approach and Literature Review

Dona Pickard

Abstract In this chapter we develop the conceptual links between urban agriculture and quality of life in order to provide a tool to evaluate whether and how UA can improve urban communities' well-being. We present the logic of selecting the specific indicators of urban quality of life which we use in our research. These are the QoL dimensions that are most commonly affected by UA activities, according to the literature: cooperation, trust and social cohesion, education, environmental awareness and sustainable consumption, work satisfaction, leisure, physical and mental health, enhanced biodiversity and pleasant public spaces, economic opportunities for local business. Bearing in mind the distinctions between the opportunities and real outcomes of a good life, as well as the external conditions and the individual dimensions of QoL (Veenhoven, 2000), we work with indicators at the individual, local community and city level. Economic benefits, education and environmental awareness are most often studied at the individual and household level; social cohesion and integration at the local community level; environmental quality at the local neighbourhood and city level. We measure these indicators using both self-reporting and unbiased data analyses.

Case studies from around the world, as well as theoretical writings on the processes that characterize UA, have accumulated evidence that UA practices have various beneficial effects on urban communities. Possible negative effects of UA have also been studied, and they include risky practices of re-using urban waste in urban and peri-urban agriculture (Drechsel et al., 2015; Valipour & Singh, 2016), possibility of contamination of the produce by chemical pollution of the soil (Gerster-Bentaya, 2015), as well as critique of urban agriculture projects leading to middle-class urban gentrification (McCintock, 2013). Keeping in mind that UA does not always impact communities and urban spaces positively, we focus our study on the mechanisms through which it can improve the well-being of urban dwellers. The positive impacts have been linked to improved social, economic, environmental and spatial quality of

D. Pickard (✉)
Institute of Philosophy and Sociology, Bulgarian Academy of Sciences, Sofia, Bulgaria

© The Author(s), under exclusive license to Springer Nature Switzerland AG 2022
D. Pickard (ed.), *Urban Agriculture for Improving the Quality of Life*, Urban Agriculture, https://doi.org/10.1007/978-3-030-94743-9_3

urban life. Still, few studies have addressed UA's impact on local communities from an interdisciplinary point of view (Lohrberg et al., 2016). Such an approach was adopted in our research in order to gain a wider picture of the set of mechanisms through which UA could improve urban communities' well-being, including social, economic, spatial and environmental means of affecting individuals' and communities' welfare. It was our aim to test to what extent this dominant positive narrative reflects the reality of UA practices in Sofia and whether it fits the potential of the capital city of Bulgaria to improve the quality of life of urban communities through such practices. This is why we narrowed down the scope of what we mean by urban agriculture and urban gardening to *the practices of growing plants and keeping animals for food in urban and peri-urban environments in and around the City of Sofia, which, apart from producing food, may lead to a range of social, economic and environmental benefits.* As the focus of our study was the link between all types of urban and peri-urban agriculture practices and their effects on individual and community quality of life, our definition encompasses all dimensions studied in the literature: the type of economic activity (for self-sustenance, hobby, charity or business), the categories of produced food (unprocessed fruit and vegetables, processed goods, honey, etc.), where exactly in relation to the city centre where the practices take place (intra-/peri-urban areas), types of production systems (hydroponics, vertical gardening, etc.), social categories of those practising UA (social and professional status, gender, etc.) or the scale of production (from industrial to balcony gardening). These dimensions have been the focus of the definitions provided in the most quoted urban agriculture sources (Van Veenhuizen, 2006: 2,[1] Mougeot, 2000:10–11,[2] Lohrberg et al., 2016: 21,[3] FAO, 2006[4]). Still, the focus on UA's impact on quality of life is what was at the centre of our research, therefore the framing of that link remains our main conceptual task.

We chose to relate the effects of UA to quality of life as, regardless of its multifaceted interpretations, it is an established concept to reflect a broad category

---

[1]"Urban agriculture can be defined as the growing of plants and the raising of animals for food and other uses within and around cities and towns, and related activities such as the production and delivery of inputs, and the processing and marketing of products. Urban Agriculture is located within or on the fringe of a city and comprises of a variety of production systems, ranging from subsistence production and processing at household level to fully commercialized agriculture."

[2]"The lead feature of UA which distinguishes it from RA is its integration into the urban economic and ecological system/.../ UA is an industry located within (intra-urban) or on the fringe (peri-urban) of a town, a city or a metropolis, which grows or raises, processes and distributes a diversity of food and non-food products, (re-)using largely human and material resources, products and services found in and around that urban area, and in turn supplying human and material resources, products and services largely to that urban area."

[3]"Urban agriculture spans all actors, communities, activities, places and economies that focus on biological production in a spatial context, which – according to local standards – is categorized as 'urban'. Urban agriculture takes place in intra- and peri-urban areas, and one of its key characteristics is that it is more deeply integrated in the urban system compared to other agriculture. Urban agriculture is structurally embedded in the urban fabric; it is integrated into the social and cultural life, the economics, and the metabolism of the city."

[4]"Urban agriculture is defined as food production that occurs within the confines of cities. Such production takes place in backyards, on rooftops, in community vegetable and fruit gardens and on unused or public spaces. It includes commercial operations that produce food in greenhouses and on open spaces, but is more often small-scale and scattered around the city."

of dimensions used to describe both the objective and the subjective aspects of well-being, as reflected in the well-known 'objective-subjective' classification by Zapf (1984), which was further developed and elaborated by Ruut Veenhoven (2000). Throughout our study, we use the term quality of life as a conceptual umbrella that reflects well-being and answers the question how good life is for individuals and communities.

Quality of life (QoL) as a concept in social sciences has so many interpretations that we had to limit its use to those indicators of urban life that have already shown correlation with UA practices in previous studies, and also ones that we could feasibly study in Sofia. That is why, while QoL researchers commonly insist on the comprehensiveness of indicators used to assess QoL and only work with it as an integral indicator of the conditions, achievements and success of the individual, his or her family, the community and society, as well as of the performance of the different levels of governance (Tilkidzhiev, 2009), we could only employ a limited number of sub-indicators of QoL that reflect the diversity of expected UA impacts on the individual, on the local community and on the city environment as a whole.

Considering the vast array of interpretations and uses of the quality of life concept in literature and the hundreds of measures to analyse them, depending on the specific goal and the object of the study (Veenhoven, 2000; Tov & Diener, 2007), we have narrowed our conceptual understanding of the link between UA and QoL to those measures of QoL which correspond to Ruut Veenhoven's four-way QoL classification matrix (Veenhoven, 2000). This analytical tool has been useful in our study because it is applicable to interdisciplinary research and it reflects the relevance of concepts from social studies, economics and biology, which are applied in our sociological, economic, environmental and spatial analyses. Veenhoven's matrix allows us to clearly define whose life we are studying at different levels of the analysis (that of the individual or that of a particular group), and what type of quality – objective or subjective – we are measuring.

In short, this matrix classifies four types of quality of life, according to two dichotomies: between potential for good life and good life itself ("potentiality v/s actuality") and between outer and inner QoL.

The potentiality/actuality dichotomy, also referred to as life chances/life results distinction, refers to the fact that "opportunities and outcomes are related, but are certainly not the same. Chances can fail to be realized, due to stupidity or bad luck. Conversely, people sometimes make much of their life in spite of poor opportunities" (ibid.: 4). An example of this is the difference between the quality of available public healthcare system and the health status of a nation. This distinction helps point research in the direction where quality is to be found – in our study we focus on those dimensions of life chances created by UA, but also on particular positive life results for individuals, communities and citizens as a whole, which have been linked to UA in the literature.

The second dichotomy in Veenhoven's matrix is between outer (external) and inner (internal) QoL. This distinction refers to the difference between whether a value refers to the environment or to the individual. Thus, both life chances and life results can be characteristic either of the surrounding context or of the individual human being.

Examples of QoL dimensions relevant to UA are presented in the cells of Veenhoven's matrix (Table 3.1) which illustrates how the two dichotomies relate

Table 3.1 Quality of life dimensions relevant to urban agriculture

	Outer qualities	Inner qualities
Life chances	Access to education Freedom of expression and social and economic action Equality Social security Clean air, safe drinking water, moderate climate Etc.	Physical and mental health Schooling Social skills Available capital Etc.
Life results	Ability to rear children Caring for friends and relatives Being an active citizen contributing to the community's goals Etc.	Satisfaction with one's job, relationships, and living environment Overall contentment with life Etc.

to each other. Along the life-chances row, external qualities of life chances relate to social, economic, environmental and other conditions for thriving, such as access to education, freedom, equality, social security, clean air and drinking water, etc. In the same row, internal qualities of life chances refer to the capacity of the individual to make the best of the external opportunities available to them. This capacity includes physical and mental health, schooling, social skills, available capital, etc. Along the life results row, external value of a good life relates to how useful one's well-being is to others and the environment: ability to rear children and care for friends and relatives, to be an active citizen contributing to the community's goals, etc. The internal life results are manifested in the appraisal of life-aspects, for example satisfaction with one's job, relationships, and living environment, as well as in overall attitude and contentment with life. All these QoL dimensions are dynamic and they are subject to change, hence they can be influenced by UA practices. The links between these QoL dimensions and urban agriculture practices are further described below.

Looking at the effects of UA on the various dimensions of QoL through Veenhoven's matrix allows us to easily differentiate between direct effects of UA on well-being (when urban agriculture practices lead to positive life results), and those effects that improve the individual's or community's chances to have a good life. In the case of Sofia, the parallel between potential and already manifested benefits from UA is of particular relevance, as existing practices that are specifically aimed at improving different dimensions of well-being are still few, therefore it is challenging to assess the impact of their practical "results" on the community and individuals. On the other hand, when civil activists and local authorities discuss whether to stimulate UA activities and how, the discussions are mainly focused on the assessment of the potential of UA to provide desired outcomes to the citizens of the capital. Therefore, governance decisions should be based on reliable data on which elements of that potential could be expected to yield the broadest set of well-being results.

The outer-inner quality dichotomy is also very relevant to our study, as we recognize that UA can have different impacts on practising individuals when compared to neighbouring communities. Additionally, the effect it can have on the city environment is also very important and although it could be classified as a representation of the outer qualities in Veenhoven's matrix, we have a special focus on it, as it is particularly relevant for our spatial and environmental analyses. This is why we approach the analysis of how QoL chances and results are affected by UA at three levels – that of the individual, that of particular communities, and that of the city as a whole. Below, we break down what specific links we have been looking for in our research at each of these levels, based on previous literature findings, after which we describe in brief how we approach this task methodologically in each of the four analyses – social, economic, spatial and environmental.

3.1 Levels of Analysis

3.1.1 Individual and Household Level

At this level we analyse how urban agriculture affects the individuals that partake in UA activities, as well as their immediate families and households. In much of the literature, life chances and life results from UA practices are not distinguished and are analysed as synonymous, just as outer and inner qualities. The distinctive benefits of urban agriculture for individuals and families have often been analysed broadly within the realm of objective dimensions of well-being and living standards, but also have been linked to the subjective and dynamic self-appreciation of the state of being of an individual or that of a social group in physical, psychological and social aspects.

Comprehensive literature reviews on the benefits of UA present a large scope of effects that cover both the opportunities that UA creates for a good life for individuals and their families, such as access to good quality produce, including for those family members who don't garden, but also increased emotional health scores among the elderly and improved physical activity among youth and adults (Draper & Freedman, 2010); additionally, urban gardens have been shown to deepen social ties, creating opportunities for improving the social capital of individuals, but research has shown that this also leads to concrete positive results – a stronger overall sense of community belonging (Prové, 2018). Similar findings about the most widespread positive effects of UA on gardeners related to improved diets and access to fresh fruit and vegetables and better social bonding across class and status is found by Veen and Eiter (2018). They also point out another widely explored benefit of UA, namely the importance of urban gardening to disadvantaged people, whose social networks rely particularly strongly on taking part in community gardening activities (ibid.: 153).

Similarly, the therapeutic effects of gardening have been analysed under the broad umbrella of UA benefits. Simon-Rojo et al. (2016) describe therapeutic gardens as having "basic healing effects" that "can be used for the treatment of mental disorders, autism, Alzheimer's disease or cerebral paralysis, addiction to drugs and alcohol, etc.", and when used for neuron-rehabilitation, they "stimulate the sight, smell and touch, awakening the senses and thus with them memories and emotions" (Simon-Rojo et al., 2016: 25). Tara Garnett (2000) also describes how UA can contribute to improving physical and mental health by providing the opportunity to consume larger amounts of fresh food and access to activities that are expected to boost physical health and relieve stress.

Another dimension of the chances for good life that is explored when studying UA is food security for individuals and households at risk. Physical and economic access to food (proximity of grocery shops, markets, food aid, etc. and the financial capacity to purchase food, respectively), are discussed at length by Maria Gerster-Bentaya (2015) who also states that poor nutrition, apart from directly being blamed for negative health outcomes, such as anaemia, cancer, diabetes, and obesity, "determines a person's well-being as a whole" (Gerster-Bentaya, 2015: 143). Apart from better access to perishable fresh food for disadvantaged individuals and families, growing vegetables and fruit in the city has been shown to improve food security by freeing up income for other, more costly foods, such as meat and dairy (Prain & Dubbeling, 2011 cited in Gerster-Bentaya, 2015).

Furthermore, producing food for sale in urban settings (commercially or selling extra produce), contributes to the income of families, especially considering the fact that UA can be combined with other occupations, thus multiplying the sources of income, although research from around the world has shown that the wealthier the producers are, the higher economic benefit they have from commercial UA activities as they have better access to land and means of production (Gerster-Bentaya, 2015). The economic benefit that UA can bring to individuals and households is a measurable indicator relevant to the inner qualities of the life chances one possesses. Moustier and Danso (2006), quoted in Dubbeling et al. (2010: 10), have found that despite the fact that urban food producers in the Global South are often small and their production levels and turnover are not high, altogether UA creates employment and income for many poor urban households and contributes to incomes that are at least equivalent or higher than the official minimum wage.

While the recreational opportunities that UA practices provide are often viewed as a QoL value on community or neighbourhood level (Draper & Freedman, 2010; Sun, 2005) it also has a salient personal aspect for the individual when it contributes to stress relief and improving one's physical and mental health through productive and recharging activities. This is highlighted by Pourias et al. (2016), who specify that the leisure benefits associated with UA may not only be associated with the gardening activities themselves, but with the garden space as a whole, where a person can enjoy reading, playing, eating outdoors, etc. Urban agriculture also impacts the outdoor space that a family or household shares in a family garden. Private productive green spaces and adjacent leisure areas for entertaining or relaxing, or the combination of the two (a seating and eating area under a vine, for example), create space for socializing and bonding, which adds to the social benefits of UA.

When it comes to life results, the QoL dimensions that are seen by researchers as most relevant to the effects of UA, are related to the personal satisfaction gardeners develop from being able to cope with life challenges. This satisfaction may come from appreciating the development of new gardening skills, but also academic and research skills and being able to work responsibly in a team for young people and a feeling of appreciation for their work by experienced adults (Krasny & Doyle, 2002). The personal feeling of empowerment that comes with building communities around gardening and farming activities and being able to help the producers of one's food is also an inner dimension of the QoL outcomes that has been identified in previous research (Sumner et al., 2010). While difficult to measure, this dimension is often the central focus of both literature on QoL and on UA. The satisfaction that individuals experience with their self-perceived capacity to overcome life challenges, regardless of their social, economic or health status, is a key indicator of their personal quality of life (El Din et al., 2012; Tilkidzhiev, 2009). This satisfaction and feeling of empowerment at an individual level is directly linked to the belief in one's ability to exert control over and direct one's life, including one's education, employment, relationships, etc. and it is also directly linked to heightened levels of civic participation, helping and empowering others (Tilkidzhiev, 2009), so it is generally expected to contribute to more equal communities and societies. These links between the analytical levels of QoL dimensions are useful to make extrapolations to dimensions that are more difficult to measure, as long as the methodological difference between measuring personal satisfaction with life and extrapolating how it could impact on community and city-level quality of life is strictly defined.

The listed QoL indicators that are most often linked to practising UA at individual and household level are in line with the conviction that has been reflected in scientific approaches to well-being since the 1970s: that "welfare is more than just material wealth". It spans a much wider range of individual and social characteristics that could be summarized as "people's chances for leading long, healthy and creative lives" (Veenhoven, 2000: 18-19).

3.1.2 Community Level

At the community level, we analyse the effects of UA on a broader level that is above the individual citizen or household, but not as general as the city level, which refers to the social, economic and ecological processes that characterize the urban space, its inhabitants and governance as a whole. By community we mean any social group that could be characterized by a commonality that creates in its members a sense of belonging to the group. In our research, we base our understanding of the different types of urban communities on the conceptual frameworks provided by one of the leading community sociologists in Bulgaria, Nikolay Tilkidzhiev (Tilkidzhiev, 2009), by the renowned social theorist Ferdinand Tönnies who introduced the community-society (Gemeinschaft-Gesellschaft) dichotomy (Tönnies, 1988), and by Étienne Wenger (Wenger, 1999) who researched the importance of "communities of practice" for the process of social learning. They all share an understanding of communities as social groups with members connected by a number of commonalities. Such commonalities could be shared values, interests, shared history, territory or activity. More specifically, we use Tilkidzhiev's definition of community. It defines community as a form of "*coexistence* of people, norms and affiliation, in which people participate in common, joint *activities*, in which they communicate and which connect them together", as well as "any aggregate of human beings, who come into social contact, where this aggregate is created and exists on the basis of at least one socially significant characteristic related to the execution of one, several or a system of *social action*" (Tilkidzhiev, 2002: 28-31). It is the focus on action and shared living experience which is so relevant to our study of the phenomenon of UA, as it is strictly linked to a particular activity and the changes in the urban environment that it leads to, as well as to the experiences and their effects on the QoL that it brings to the people involved in it.

Following this definition, we can identify urban communities that exist in many different types and forms, but we narrowed our research to two types that are most relevant to the study of urban agriculture and the quality of life – the communities of those practising UA, and the communities of people who are directly influenced by UA activities. These could be territorial communities in the close neighbourhoods of UA activities, as well as the people from the immediate circles of those practising UA, but not members of their households. Communication and interaction are immanent characteristics of these communities, as well as sharing certain space and time. Along with these, there are also virtual online UA communities sharing and supporting the idea of UA across the country and beyond, which do not share all these characteristics. We did not study them due to the difficulty of involving them in interviews and large-scale representative surveys, as well as the uncertainty as to the extent to which they are involved in UA practically.

In our work, therefore, the urban communities that are our object of our study can be defined as functional groups that are territorially bound to an urban settlement and are involved in a social action that works as a founding base of the group – whether it is practising UA themselves or experiencing a direct effect of UA activities. These could be neighbourhood groups, UA practising groups of like-minded people with shared values regarding sustainable living, formal and informal groups like practising commercial producers, processors and traders, including farmers' markets organizations and groups for direct sales, various associations working in the field of sustainable development and urban planning, who support UA activities and policies. When applying this understanding of urban communities, we take the viewpoint of both social scientists and urban geographers whose definitions of community differ in that the former associate communities with a group of people characterized by close and direct social ties who share "common interests or social concerns", while the latter focus more on belonging to a specific limited area (Sun, 2005: 5).

Analysing the impact of UA on the quality of life dimensions for these types of communities, we make use of the "community quality of life" concept used by Yinshe Sun (2005): this focuses on the community environment, resources and services, which the members of the community believe have an effect on their individual quality of life (ibid.). As mentioned above, the individual experiences of personal satisfaction within a community are linked back to feelings of closeness and increasing social cohesion and social integration. These could be classified as outer qualities of the life chances according to Veenhoven's classification. As such, they provide opportunities for social inclusion of the individuals through encouragement to participate in social life, according to the principles of voluntary action and equal access.

Practising urban agriculture in a voluntary manner leads to new contacts or broadening and strengthening already existing ones. This is how UA networks that reflect social cohesion appear – they contribute not only to supporting the specific network activities and members, but to the general democratic culture, where "actors may continue to pursue 'their own thing' while communicating and cooperating with other like-minded actors who are 'doing other things' helpful in moving towards local, social and ecological sustainability in food and all other matters" (Walliman, 2015: 85). Thus, social cohesion is central for community QoL as it provides "better life for all" (Cloete & Kotze, 2009: 14).

Research into the possible ways UA contributes to community building and stronger community ties has not yet definitively proven that involving citizens in agricultural activities in or around cities necessarily results in an improved community spirit which in turn can be interpreted as outer quality of life chances and life results (in the first case as an improved social environment allowing for more equal and active social engagement, and in the second as active civil participation). There is plenty of research that points to the capacity of UA to enhance social inclusion of different social groups such as minorities, the disabled, women and youth at risk of poverty (Gonzales Novo & Murphy, 2000; Dubbeling et al., 2010). Sumner et al. (2010) analyse both research on Community supported agriculture (CSA) and a peri-urban CSA case in Ontario, USA, to conclude that although not all participants in the initiative develop a community identity, these types of activities provide a fruitful ground for germinating movements seeking healthy food and following a desire to support local farmers and generally a more sustainable food system. In our study we raise the question of whether urban agriculture activities improve the social networks of individuals or it is the fact that these individuals are generally better networked and socially active that motivates them to partake in UA, but the literature has shown that regardless of the QoL results, UA at least offers a QoL chance for "a community-building connection with farmers, neighbours and landscapes" (Tegtmeier & Duffy, 2005, quoted in Sumner, Mair and Nelson (2010: 59).

Two ways that the literature has systematically shown how UA can contribute to the social inclusion of disadvantaged individuals and social groups are through their involvement in activities that change the social profile of public spaces and through the mechanisms of the solidarity economy. In the first case, groups that are traditionally excluded from community life, especially because of their low-income profile, are allowed to and welcomed into public spaces where they can establish and get involved in social networks and become socially empowered despite their monetary resource deficits (Slater, 2001). The engagement in UA activities in public spaces is important to community building as it has shown it can boost the participants' collective identity through a shared relation to a place, as well as shared responsibilities and interests (Koopmans et al., 2017: 162). In the second case, non-financial resources, such as time, materials and specific skills, are harnessed in producing and exchanging food and food-related products and services to pool resources for mutual use and in this way provide long-term "existential security for an ever-increasing number of people" (Walliman, 2015: 80). Although "it takes time and some skills to participate in such exchange circles", the solidarity economy around UA has been demonstrated to build communities around alternative economic, cultural and democratic opportunities created in a niche economy (ibid.).

A dimension of neighbourhood quality of life related to UA, but not based on personal involvement in UA activities, is the satisfaction with the neighbourhood. As demonstrated by El Din and colleagues, the most important factors that determine individuals' satisfaction with neighbourhood and community life are the quality of social interactions, the levels of neighbourhood crime and public facilities, such as parks and other public green spaces (2012: 88). All of these could be impacted positively by UA. As already pointed out, research has shown how such activities can improve social relations, and as to crime, studies that have outlined such an effect enhances the positive effect UA has on social control and social cohesion in neighbourhoods (Prové, 2018). Regarding improved experiences of public spaces, UA contributes to the aestheticization of abandoned neighbourhood plots and makes interaction with green spaces more meaningful when they are linked to a productive activity (ibid.).

3.1.3   City Level

As mentioned above, the typology of social relations that differentiates between community and society (Gemeinschaft and Gesellschaft) was introduced by F. Tönnies (1988), who based the dichotomy on the nature of the dependencies that characterize these two types of social structures. While the interdependencies in the community are commonly limited to a system that is self-sufficient to a large extent and based on immediate social ties and personal participation, social relations in a society are built upon the principles of complementarity and mediated relations, based on rationality, division of labour and delegated power. We use this differentiation to separate more clearly the different effects that UA might have on the urban environment and society as a whole, as compared to its effect on local urban communities.

At the city level we investigate mainly spatial and environmental factors that characterize the potential of urban space to provide a more liveable space and a better life for its inhabitants. We also look at social and economic processes on a larger scale, such as governance issues and the economic environment which can be influenced by the processes of UA development. In most cases, the urban dimensions of QoL related to UA are linked more to the opportunities for a good life rather than good life results.

One of the central benefits of UA to the city environment is claimed to be the ecosystem services provided, such as reduced CO_2 emissions of the food system due to sourcing locally and using less packaging, reduction of run-off water and storing it in soils and technosoils when it is used for UA irrigation and improved urban microclimate due to plant transpiration (Tornaghi et al., 2016: 166). Still, there is research that points to the many variables that influence these positive effects (Prové, 2018: 47), the most important being the scale at which UA is practised in a city. As Sofia is still a modest example of UA development, and the few business and community urban agriculture activities are too insignificant to cause a measurable effect on the city food system carbon footprint or average temperatures, we will be investigating the characteristics of these initiatives at micro-level, analysing their biodiversity, use of water and recycling resources, so we can establish the potential of similar activities to improve the urban environment and make its use more sustainable. Related to the environmental effects of UA is food awareness and the more sustainable consumption models that are fostered by participating in UA activities (Sumner et al., 2010), which in the long run could create a more wide-spread culture of sustainable living.

Another effect of UA for cities widely explored in research is the improvement of public spaces that allows for more social inclusivity and offer citizens chances to communicate, co-create and engage not only in UA but in public life as a whole. Because of the contested use of land in rapidly growing cities, UA stands at the crux of the values competition that regulate what urban space is used for. In most cases, the dominant value that rules this competition is the economic return that each space can yield (Wirth, 2016: 119, Purcell, 2002), and less tangible benefits such as human happiness, are not a priority. Looking at QoL though, while economic return is only a prerequisite and gives a chance for higher quality of life, and with urban sprawl – only a few citizens are affected by it – happiness is a quality of life result.

This is why we put a special focus on the spatial transformations that UA leads to which have been shown to affect citizens' appraisal of life positively. Here, human interaction in a shared space leads to happiness (Gehl, 2016: 608) and even more so when this space is not commercialized or dedicated to "passive spectoralism", but more closely integrated with natural spaces and active participation (Wirth, 2016). As UA offers the opportunity for human interaction in a shared space and within a more natural environment than the typical built-up city areas offer, we look at opportunities to integrate such activities within the existing spatial resources of the city to make the urban environment healthier and more pleasurable.

Having discussed the non-monetary benefits of UA, it must also be said that there are also aspects of UA that might boost the urban economy as a whole and lead to a higher income for the city budget. As UA can improve the economic standing of households and small businesses, it can also influence the city economy if performed at a larger scale that can lead to the emergence of a chain of support businesses, such as the seeds and fertilizers trade, consultation services, etc. When UA is developed at a commercial scale in cities, as in many cases around the world, it has demonstrated higher growth rates than rural agriculture, mainly due to its comparative advantage due to its proximity to urban consumers and lower transport and cooling costs, especially important for perishable goods (Dubbeling et al., 2010: 12). The potential of UA to be the foundation for a city-wide solidarity economy should also not be neglected. Although city environments do not commonly encourage alternative economic systems (Wirth, 2016), widespread urban agriculture activities have been shown to create fruitful ground for an economic life in which citizens have a basis of subsistence to rely upon. This is the solidarity economy that UA can create, encouraging self-employment and sustainable relations of trust, exchange and mutual help (Walliman, 2015).

Finally, the social dimensions of urban quality of life at city level are also of interest to our research. They are impossible to analyse at the level of deep social ties as "the contacts of the city may indeed be face-to-face, but are nevertheless impersonal, superficial, transitory, and segmental" (Wirth, 2016: 118). That is why we measure them through other social indicators, namely general satisfaction with life in the city and good governance, based on active civic participation. Research has shown that evaluation of the quality of the environment and the management of environmental problems is one of the significant predictors of satisfaction with life at the city and regional level, especially for young people (El Din et al., 2012: 88). As this satisfaction is often linked to identifying with a place through engaging in group activities, it could be deduced that UA not only contributes to being more satisfied with one's neighbourhood environment, but also with one's city. There is no clear cause and effect link between these two indicators; our empirical study aims to look more closely at possible ways they can influence each other. Satisfaction with city life is also linked to a perception of good city governance, especially when one participates in it as a civil actor. Good governance characterized by active civic participation which often accompanies UA is linked to good urban life. This is because of the experience of participation leads to a feeling of living in an integrated society, which provides existential security. As opposed to the feeling of being lost in a city, where "the bonds of kinship, of neighbourliness, and the sentiments arising out of living together for generations under a common folk tradition are likely to be absent or, at best, relatively weak" (Wirth, 2016: 118), being engaged in civil action, including through UA, can empower citizens to identify with their cities and with their fellow citizens (Pickard, 2018).

To summarize, we study physical and mental well-being (including the feeling of empowerment) and recreation, as well as economic benefits, at the individual and household level; cooperation, social cohesion and integration, as well as greater satisfaction with one's living environment at the local community level; and environmental quality, including public spaces, good governance and the potential for boosting the city economy at the city level. Education and environmental awareness are relevant to all three levels, but especially to the individual and city levels. We measure these indicators using both self-reporting and unbiased data analyses.

Having established the QoL dimensions that we will be examining in relation to UA practices in Sofia, the next step in our work was to decide on the indicators that we would use to measure the effect of these practices. We used an interdisciplinary approach to this task where the design of the research instruments and methods, and the analysis of the empirical data generated was developed in parallel in three sub-studies, these being socio-economic, environmental and spatial in nature. They developed their own set of indicators reflecting the links with urban agriculture that are most prominent in the particular research domain. These are presented in detail in each of Chaps. 4, 5, 6 and 7.

Depending on the methodological toolkit of each discipline, each sub-study relied more on objective indicators or more on subjective indicators of QoL. Objective indicators derived from representative statistical databases, nature observation and monitoring were used wherever appropriate and whenever they were available. Although reliable, such data are for the most part not available or difficult to obtain, so subjective indicators relating to the personal perceptions of one's own well-being and that of the community or the urban society as a whole, were also used. Although it is more difficult to design instruments to reliably test these subjective interpretations of one's well-being, subjective indicators measured from the point of view of the individual are just as important for our research, as they are the closest reflection of the inner life results as located in Veenhoven's matrix.

The socio-economic study was based on a combination of methods. It started off with desk research of the range of positive effects of UA on individuals, families and local communities, as well as of the existing practices of UA on the territory of the Municipality of Sofia. The desktop study also reviewed all municipal strategies and policy programmes that aim to improve the quality of life of the citizens of the capital in order to identify local policy goals that correspond to the effects of UA as expected in the literature. We also reviewed these policy documents for specific mention of UA or UA regulation to confirm our previous knowledge that there is no recognition of the phenomenon in the municipality's strategies and regulations.

The second part of the socio-economic study employed qualitative methods of collecting data. We interviewed 21 urban agriculture actors from the city, representative of all types of stakeholders in the field: a representative of the Sofia City Council, a chairperson of the standing committee on the environment, agriculture and forestry; a representative of the architecture department of a district administration (a smaller territorial and administrative entity within Sofia Municipality) – a partner in an international project who developed a model for an agri-business incubator in Sofia Municipality which was never applied in practice; a representative of a municipal company which offered free land without a written contract for a community garden in Sofia; representatives of environmental NGOs who work in the field of UA development; urban agriculture practitioners in community gardens, citizens who live in close proximity to urban vegetable gardens; representatives of research institutions who develop UA projects; commercial urban farmers; organizers of weekly farmers markets with local produce; managers of two restaurants using local produce; an activist from civil groups which are attempting to help disadvantaged groups to become reintegrated in society by involving them in UA activities, actors engaged in creating and running vegetable gardens in two schools in Sofia and the directors of the urban experimental field of the Forestry University in Sofia, which manages over 75 acres of urban land for farm animals, fodder, orchards, vegetable plots and a greenhouse, vineyards, an apiary and a dendrarium.

Expert statements on specifically targeted topics were also collected. These included a detailed description of the land resources available for UA development and expert interpretation of how the existing legal framework in Sofia allows for the establishment and spread of UA activities by a renowned architect with a long-lasting interest in the UA potential of Sofia and by a landscape architect who is a leading expert on "Green systems, recreation and ecology" from the municipal urban planning company. An expert statement on the potential of UA to be integrated in school curricula and the environmental education of children in schools and kinder-gartens was obtained from a senior expert in natural sciences from the regional education administration. Lastly, a statement on the way the municipality works with citizens and bottom-up UA organizations to negotiate possible ways of UA devel-opment in the city was obtained from an expert from the municipal foundation "Sofia Development Association" which has the aim of encouraging and maintaining a dialogue between civil society, academic institutions, businesses and the municipality.

To put up for discussion and receive feedback on the preliminary findings from the interviews and the expert statements, five focus groups were conducted with different types of stakeholders: with gardeners from a community urban garden; with citizens living near a UA practice; with representatives of school administrations, and members of school boards of schools that offer gardening to their students; a focus group with the staff of the "Green system, ecology and land use" department of the Environment Directorate at Sofia Municipality. In order to enable cross-sectoral dialogue, we organized a fifth focus group that brought together representatives from all types of stakeholders from the previous four groups.

The main areas of discussion covered in this qualitative stage of the study included the following topics:

- general awareness of urban agriculture as a phenomenon, including perceived beneficial effects on the QoL of urban citizens and communities;
- awareness of the types of actors that are involved in the development of UA in Sofia and how each respondent relates to the rest of the actors/types of stakeholders;
- awareness of the existing land, water and resources that are or can be used for UA in Sofia, as well as opinions about the social groups that could benefit from being involved in UA activities;
- the goals of the UA initiatives when the respondent is involved in them, as well as an evaluation of the extent to which those goals have been achieved (including the social, economic, environmental and spatial goals);
- the criteria used by the respondents to evaluate to what extent their goals regarding UA are being achieved or addressed adequately;
- general satisfaction with one's life.

In the final stage of this study, we carried out a survey of the practices of UA and attitudes towards them among the residents of three districts in Sofia – a central one (Triaditsa); a residential district where the predominant housing type is the prefabricated concrete panel buildings with green spaces between them (Mladost), and a peri-urban district (Pancharevo). The study was representative of each of the three districts. These were selected based on an analysis of the potential for urban agriculture in Sofia Municipality; and one of the objectives of the study was to examine the attitudes and opinions of Sofia residents living in three areas of diverse characteristics, to determine to what extent the people living there have a specific attitude to the very idea of UA and different attitudes to getting involved in UA practices, as well as their different personal experiences.

The specific data that the study aimed to gather with a set of 70 questions covered the following topics:

- the level of awareness about UA as a phenomenon of the inhabitants of the three districts under study;
- people's opinions on the challenges and opportunities UA faces in general, as well as how tolerant they would be if UA was developed in their own area and in the city as a whole;
- to establish what share of people practise UA themselves, and what drives them to do so;
- the attitudes that non-UA-practising citizens have towards UA and their readiness to join existing practices or start one themselves;
- to assess people's satisfaction with the urban environment they live in, including spatial and architectural characteristics and the social contacts they have in their neighbourhood.

Although not representative of the whole city, the survey was designed to reflect the territorial diversity of its population in terms of social and economic status, lifestyle and typical access to and use of green spaces. The survey was administered in the form of a standardized face-to-face interview. The sample included 750 citizens altogether; the share of respondents from each of the three Sofia districts corresponded to the ratio between their population numbers. The survey sample was constructed as a stratified two-stage cluster sample. Within each of the three districts (central, residential and peri-urban) 24, 39 and 11 clusters of residential addresses were sampled along randomly selected streets. Each cluster consisted of 10 or 11 respondents who were selected by the interviewers according to a scheme that ensures gender and age ratios of the sample reflecting the ratios in the general population of the City of Sofia.

The environmental study was based on three approaches in order to study aspects of urban agriculture impact on environmental components of the City of Sofia. The quality of water used for irrigation, the content of heavy metals in soils and local produce were evaluated through laboratory tests. Initial observations of the biodiversity in three UA sites in Sofia covered not only the positive impact but some of the main risks that should be considered while planning and developing UA in order to ensure human health and environmental protection. A brief literature review of UA impact on other environmental factors was also conducted. As a result, an initial assessment of UA impact on selected groups of organisms (land snails, butterflies, moths and birds) was made.

The literature often points to the positive impacts of UA on reducing the effect of the urban heat island, on mitigating the destructive effects of runoff water, protecting and improving soil quality, reducing the carbon footprint, the potential contribution to improving the condition of polluted urban areas, stimulating recycling and sustainable management of organic household waste through composting. Most of these claims can only be confirmed in a larger study in an environment where UA is widespread. Since known community gardens for UA in Sofia City cover just about 1.6 acres in August 2020, these wide-ranging UA effects are analysed on the basis of a desktop study. The environmental study within the project does not include the private gardens in the urban and peri- urban areas

The spatial study included desk research of various historical, theoretical and regulatory aspects, as well as a review and analysis of the specialized literature, historical data, archive maps and drawings. This stage was followed by extensive and thorough field work, exploring the urban and architectural specifics of three very distinct types of urban environment: a classical high intensity intra-urban district, a peri-urban "bedroom community" district, and a suburban region including ten rural settlements and their adjacent territories. Several types of urban agriculture practices were traced and analysed and the data were mapped in their GPS coordinates and represented as interactive maps.

The results showed how urban agriculture practices differ depending on the specifics of the district. The statistics, graphics and maps produced proved that the centuries-old bottom-up traditions enhanced with contemporary socio-cultural practices are valid approaches for improving quality of life on individual, communal and city level, mostly affecting the actual internal and the potential external dimensions.

References

Cloete, P., & Kotze, F. (2009). *Concept paper on social cohesion/inclusion in local integrated development plans.* Department of Social Development, Republic of South Africa. Retrieved June 12, 2021, from https://docplayer.net/30508732-Concept-paper-on-social-cohesion-inclusion-in-local-integrated-development-plans.html

Draper, C., & Freedman, D. (2010). Review and analysis of the benefits, purposes, and motivations associated with community gardening in the United States. *Journal of Community Practice, 18*(4), 458–492. https://doi.org/10.1080/10705422.2010.519682

Drechsel, P., Keraita, B., Cofie, O. O., & Nikiema, J. (2015). Productive and safe use of urban organic waste and wastewater in urban food production systems in low-income countries. In H. de Zeeuw & P. Drechsel (Eds.), *Cities and agriculture: Developing resilient urban food systems* (pp. 162–191). Routledge.

Dubbeling, M., de Zeeuw, H., & van Veenhuizen, R. (2010). *Cities, poverty and food: Multistakeholder policy and planning in urban agriculture.* Practical Action Publishing.

El Din, H. S., Shalaby, A., Farouh, H. E., & Elariane, S. (2012). Principles of urban quality of life for a neighbourhood. *Housing and Building National Research Center Journal, 9*(1), 86–92. https://doi.org/10.1016/j.hbrcj.2013.02.007

FAO. (2006). *The state of Food and Agriculture* (FAO Agricultural Series (29)). Retrieved June 12, 2021, from http://www.fao.org/3/w1358e/w1358e00.htm

Garnett, T. (2000). Urban agriculture in London: Rethinking our food economy. In M. Bakker, S. Gündel, U. Sabel-Koschella, & H. de Zeeuw (Eds.), *Growing cities, growing food: Urban agriculture on the policy agenda* (pp. 477–501). German Foundation for International Development.

Gehl, J. (2016). Three types of outdoor activities, life between buildings, and outdoor activities and the quality of outdoor Space. In R. LeGates & F. Stout (Eds.), *The city reader* (pp. 608–617). Routledge.

Gerster-Bentaya, M. (2015). Urban agriculture's contributions to urban food security and nutrition. In H. de Zeeuw & P. Drechsel (Eds.), *Cities and agriculture. Developing resilient food systems* (pp. 139–161). Routledge.

Gonzales Novo, M., & Murphy, C. (2000). Urban agriculture in the city of Havana: A popular response to a crisis. In M. Bakker, S. Gündel, U. Sabel-Koschella, & H. de Zeeuw (Eds.), *Growing cities, growing food: Urban agriculture on the policy agenda* (pp. 329–348). German Foundation for International Development.

Koopmans, M. E., Keech, D., Sovova, L., & Reed, M. (2017). Urban agriculture and place-making: Narratives about place and space in Ghent. *Brno and Bristol. Moravian Geographical Reports, 25*(3), 154–165. https://doi.org/10.1515/mgr-2017-0014

Krasny, M., & Doyle, R. (2002). Participatory approaches to program development and engaging youth in research: The case of an inter-generational urban community gardening program. *Journal of Extension, 40*(5). Retrieved June 12, 2021, from https://archives.joe.org/joe/2002october/a3.php

Lohrberg, F., Lička, L., Scazzosi, L., & Timpe, A. (Eds.). (2016). *Urban agriculture Europe*. Jovis.

McCintock, N. (2013). Radical, reformist, and garden-variety neoliberal: Coming to terms with urban agriculture's contradictions. *Local Environment, 19*(2), 147–171. https://doi.org/10.1080/13549839.2012.752797

Mougeot, L. J. A. (2000). Urban agriculture: Definition, presence, potentials and risks. In M. Bakker, S. Gündel, U. Sabel-Koschella, & H. de Zeeuw (Eds.), *Growing cities, growing food: Urban agriculture on the policy agenda* (pp. 1–42). German Foundation for International Development.

Moustier, P., & Danso, G. (2006). Local economic development and marketing of urban produced food. In R. van Veenhuizen (Ed.), *Cities farming for the future. Urban agriculture for green and productive cities* (pp. 172–208). RUAF Foundation.

Pickard, D. (2018). Factors for effectiveness of social innovation in urban agriculture. An analysis of a negative case. *International Journal of Sociology of Agriculture and Food, 24*(3), 377–392. https://doi.org/10.48416/ijsaf.v24i3.4

Pourias, J., Aubry, C., & Duchemin, E. (2016). Is food a motivation for urban gardeners? Multifunctionality and the relative importance of the food function in urban collective gardens of Paris and Montreal. *Agriculture and Human Values, 33*(2), 257–273. https://doi.org/10.1007/s10460-015-9606-y

Prain, G., & Dubbeling, M. (2011). *Urban agriculture: A sustainable solution to alleviating urban poverty, addressing the food crisis and adapting to climate change*. Synthesis report on five case studies prepared for the World Bank. RUAF Foundation.

Prové C. (2018). *The politics of urban agriculture: An international exploration of governance, food systems, and environmental justice*. PhD thesis, Ghent University.

Purcell, M. (2002). Excavating Lefebvre: The right to the city and its urban politics of the inhabitant. *Geo Journal, 58*, 99–108. https://doi.org/10.1023/B:GEJO.0000010829.62237.8f

Simon-Rojo, M., Recasens, X., Callau, S., Duži, B., Eiter, S., Hernández-Jiménes, V., Kettle, P., Laviscio, R., Lohrberg, F., Pickard, D., Scazzosi, L., & Vejre, H. (2016). From urban food gardening to urban farming. In F. Lohrberg, L. Lička, L. Scazzosi, & A. Timpe (Eds.), *Urban agriculture Europe* (pp. 22–28). Jovis.

Slater, R. J. (2001). Urban agriculture, development, gender and empowerment. An alternative view. *Development Southern Africa, 18*(5), 635–650. https://doi.org/10.1080/03768350120097478

Sumner, J., Mair, H., & Nelson, E. (2010). Putting the culture back into agriculture: Civic engagement, community and the celebration of local food. *International Journal of Agricultural Sustainability, 8*(1–2), 54–61. https://doi.org/10.3763/ijas.2009.0454

Sun, Y. (2005). *Development of neighbourhood quality of life indicators*. University of Saskatchewan: Community-University Institute for Social Research. Retrieved June 12, 2021, from SunBook (usask.ca) https://cuisr.usask.ca/documents/publications/2005-2009/Development%20of%20Neighbourhood%20Quality%20of%20Life%20Indicators.pdf

Tegtmeier, E., & Duffy, M. (2005). *Community supported agriculture (CSA) in the Midwest United States. A regional characterization*. Leopold Center for Sustainable Agriculture, Iowa State University. Retrieved June 12, 2021, from https://lib.dr.iastate.edu/leopold_pubspapers/151

Tilkidzhiev, N. (2002). *Middle class and social stratification*. Lik (In Bulgarian).

Tilkidzhiev, N. (2009). Trust and well-being. In N. Tilkidzhiev & L. Dimova (Eds.), *Well-being and trust. Bulgaria in Europe? Comparative analysis upon the European Social Survey 2006/2009* (pp. 33–62). Iztok-Zapad. (In Bulgarian).

Tönnies, F. (1988). *Community and society (Gemeinschaft-Gesellschaft)* (C. P. Loomis, Trans.). Routledge (Original work published 1887).

Tornaghi, C., Sage, C., & Dehaene, M. (2016). Metabolism: Introduction. In F. Lohrberg, L. Lička, L. Scazzosi, & A. Timpe (Eds.), *Urban agriculture Europe* (pp. 166–169). Jovis.

Tov, W., & Diener, E. (2007). Culture and subjective well-being. In S. Kitayama & D. Cohen (Eds.), *Handbook of cultural psychology* (pp. 691–713). Guilford.

Valipour, M., & Singh, V. P. (2016). Global experiences on wastewater irrigation: Challenges and prospects. In B. Maheshwari, B. Thoradeniya, & V. P. Singh (Eds.), *Balanced urban development: Options and strategies for liveable cities*. Springer. https://doi.org/10.1007/978-3-319-28112-4_18

Van Veenhuizen, R. (2006). Introduction. Cities farming for the future. In R. van Veenhuizen (Ed.), *Cities farming for the future. Urban agriculture for green and productive cities* (pp. 1–18).

Veen, E., & Eiter, S. (2018). Vegetables and social relations in Norway and the Netherlands. A comparative analysis of urban allotment gardeners. *Nature and Culture, 13*(1), 135–160. https://doi.org/10.3167/nc.2018.130107

Veenhoven, R. (2000). The four qualities of life. Ordering concepts and measures of the good life. *Journal of Happiness Studies, 1*(1), 1–39.

Walliman, I. (2015). Urban agriculture as embedded in the social and solidarity economy Basel. Developing sustainable communities. In P. Stock, M. Carolan, & C. Rosin (Eds.), *Food utopias. Reimagining citizenship, ethics and community* (pp. 79–87). Routledge.

Wenger, E. (1999). *Communities of practice. Learning, meaning and identity*. Cambridge University Press.

Wirth, L. (2016). Urbanism as a way of life. In R. LeGates & F. Stout (Eds.), *The city reader* (pp. 115–123). Routledge.

Zapf, W. (1984). Individuelle Wohlfahrt: Lebensbedingungen und wahrgenommene Lebensqualität. In W. Glatzer & W. Zapf (Eds.), *Lebensqualität in der Bundesrepublik. Objektive Lebensbedingungen und subjektives Wohlbefinden* (pp. 13–26). Campus Verlag.

Part II
Dimensions of the Impacts of Urban Agriculture on the Quality of Life in Sofia: Results and Analysis

Chapter 4
Social Dimensions

Dona Pickard, Mariana Draganova, Albena Nakova,
and Emilia Chengelova

Abstract This chapter explores the potential and accomplished social effects of UA in Sofia, focusing on social cohesion and inclusion through urban agriculture practices (Sect. 4.1), the educational effects of urban agriculture (Sect. 4.2) and urban agriculture and sustainable consumption patterns (Sect. 4.3).

The analysis from the qualitative and quantitative study point to the conclusion that the respondents with the highest level of social contact in their daily living environment represent the highest share of practitioners with social motivations to practice UA and there is no substantial evidence for social inclusion that came as a result from UA activities. The educational effects of urban farming are considered as the single most valuable contribution to society that urban agriculture can have. It is a commonly shared belief that if young people know more about food and how it is produced, they would develop more responsible consumption patterns, and pro-environmental behaviour, and will make healthier food choices. Based on the interviews conducted, several patterns of sustainable consumption were derived.

4.1 Social Cohesion and Inclusion Through Urban Agriculture in the Case of the City of Sofia

As mentioned in Chap. 3, social cohesion and inclusion are the social effects of UA that are among the most commonly mentioned in the literature. To set the analysis of social cohesion and inclusion affected by UA in Sofia in a broader context, we make use of the social capital concept (Putnam, 2001, 2004; Grootaert et al., 2004; Narayan & Cassidy, 2001; Pickard, 2013). The elements of social capital are closely related to analysing QoL, as they refer to life chances – outer ones (access to education and adequate health care, clean environment, etc.) and inner ones (physical and mental health, social skills, etc.). Also, looking at social cohesion from the point of view of the encompassing frame of social capital, it is easier to bring

D. Pickard (✉) · M. Draganova · A. Nakova · E. Chengelova
Institute of Philosophy and Sociology, Bulgarian Academy of Sciences, Sofia, Bulgaria
e-mail: albena_nakova.manolova@abv.bg

together all the factors that enhance it and include them in research, so that a comprehensive picture of the interplay between UA activities and cohesion and inclusion can be constructed.

There are different dimensions of social capital that influence social cohesion (a dimension of social capital itself): the characteristics of social networks and the level of trust and solidarity in them; the shared norms and goals that allow for collective action and cooperation; access to information and free flows of information. Social cohesion and inclusion, together with empowerment and civil action are in fact the end results and areas that make the presence of social capital visible (Pickard, 2013; Grootaert et al., 2004). It is important to view social cohesion in relation to other dimensions of social capital when investigating how UA impacts on it, because due to the complex nature of social capital it is possible for UA to be very effective in producing cohesion and inclusion effects in some cases, and not in others. It is of interest not only to see what the case in Sofia is, but to also try and determine the factors for success or failure in that direction. To do this, we study all these structuring dimensions of social capital that lead to social cohesion and present them, together with other relevant dimensions of QoL, at the three analytical – individual, community and city – levels.

4.1.1 Individual and Household Level

The structure of social networks that underpin the experience of trust and solidarity, empower collective action and lay the foundations of social cohesion can only be analysed at above-individual level. Still, there is a range of inner capacities of the individual (following Veenhoven's classification) that improve the chances of each person to enter or make the best use of being a member of social networks that thrive on solidarity and cooperation and can be characterized as cohesive and inclusive. These are being physically and emotionally fit through maintaining healthy diets and physical activity, and overcoming substance addictions and the burdens of a stressful lifestyle.

In the first stage of our socio-economic study, physical and mental health was not stressed as a sought impact of gardening activities. Among the practising UA gardeners who are 26.6% of the whole sample of our second stage quantitative study, the motivation to garden in order to keep physically and mentally fit was not among the most often quoted reasons to do so, either: 15.9% started gardening in order to keep physically healthy and 21.9% to feel better mentally and psychologically (Fig. 4.1).

On the other hand, when practitioners are specifically asked to evaluate the effect of gardening on their physical and mental health,[1] the results show that this effect is very positive in most cases: 61.2% declare gardening makes them feel much

---

[1] The questions we posed in the survey regarding physical and mental health were "Do you feel healthier and more refreshed after your gardening activities?" and "Do you usually feel better mentally and psychologically after your gardening activities?"

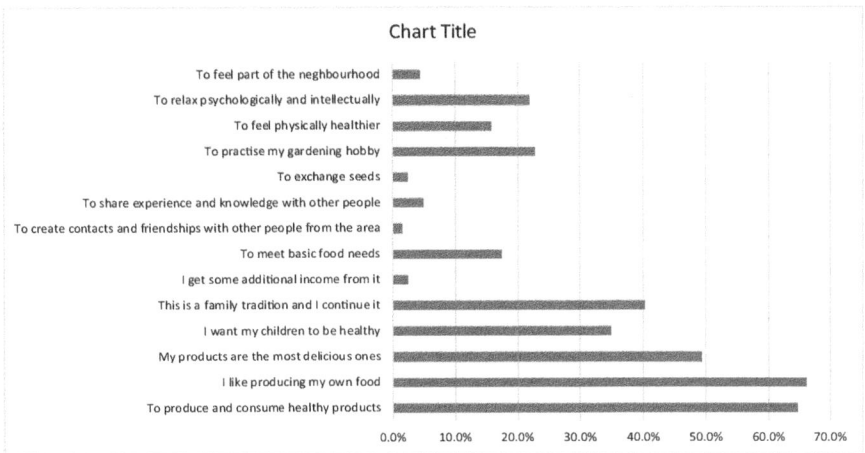

Fig. 4.1 Motivations to start urban gardening among the practitioners in the studied districts in Sofia. The total sum of answer shares is over 100 because respondents could give more than one answer

Fig. 4.2 Reported effects on physical health

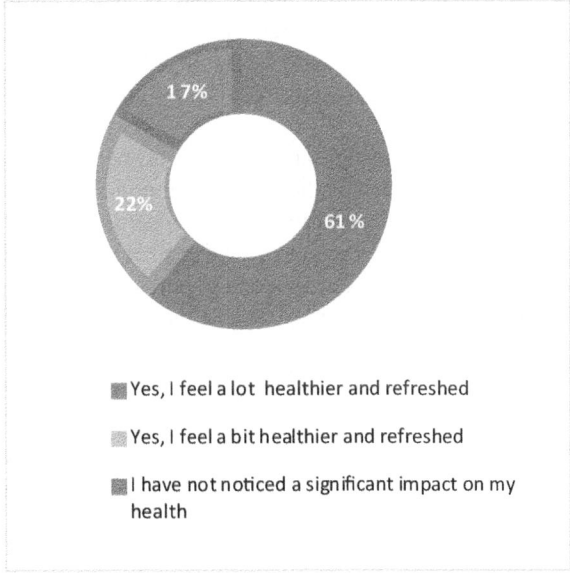

healthier physically than if they didn't garden and 76.8% report they feel much better mentally (Figs. 4.2 and 4.3).

This supports the literature which has highlighted the positive health impacts of UA (Garnett, 2000; Draper & Freedman, 2010), but also points to the fact that the effects of UA may not always be predicted and expected by the practitioners and the motivations and real outcomes of gardening do not always correspond. In the light of urban policies for improving the physical and mental health of individuals and

Fig. 4.3 Reported effects
on mental health

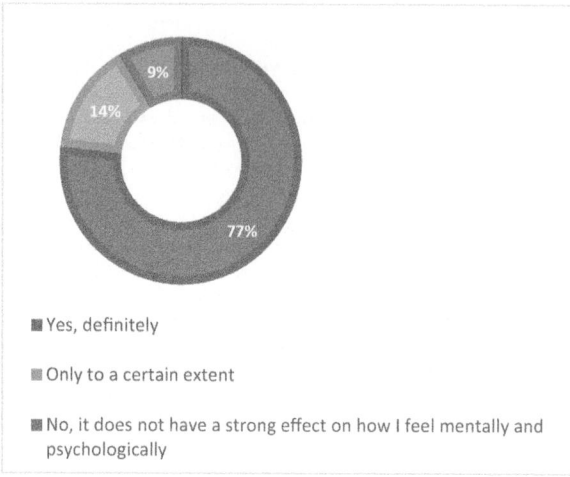

■ Yes, definitely

■ Only to a certain extent

■ No, it does not have a strong effect on how I feel mentally and
psychologically

disadvantaged groups, this is a very important detail. It justifies top-down policies that support information campaigns and awareness raising about the positive effects of UA which people may not be able to predict themselves. This is also because, while UA can benefit physical and mental health, physical and mental well-being might be a pre-condition to get involved in UA. Hence, special policies are even more necessary to encourage people with mental or physical disabilities to participate in UA.

Unlike being healthy and mentally fit, being able to provide a healthy diet for themselves and their families is a conscious goal of practising gardeners in our study. The second most quoted motivation to garden is to produce and consume healthy food (64.7%). It is difficult to establish what the share of the self-grown produce is in a gardening household, compared to purchased food, but the data from the in-depth interviews and focus groups suggest that at least for a certain period in the year and for certain vegetables, the gardeners rely entirely on what they produce and sometimes they produce so much they need to preserve it for the winter, give it away or even throw it away:

Participant 1: There's a lot of food, you can't eat it all and you have to share it. We do share. It all depends on how hard you work to produce more. For example, Stephen[2] puts in a lot of effort, he produces a lot, quite a lot.

Participant 2: I can't say that everything I have produced brings economic benefits. But for instance, it was tomatoes I produced most and from August until only one or two weeks ago I had not bought any tomatoes. And we even made something like lyutenitsa,[3] and we even threw away some.

---

[2]Names of interviewees and participants in the focus groups are changed.

[3]A relish, traditionally made as preserve for winter, made mainly of tomatoes and red peppers.

Participant 3:   Frankly speaking, since March I have not been to the market. As soon as I entered the garden, I was done (with shopping). Otherwise I used to go to the market every day. This is absolutely cool – not some economic benefit, but that (this food) is pure, not fertilized, sprayed, it is not some trash. (from a focus group with allotment gardeners)

Also, as only a very small number of respondents sell what they produce (only 2%) and most of them (81.6%) produce for their family and friends, it could be suggested that the effects from consciously producing healthy food, reaches a wider consumer group than only that of the gardeners themselves. In our research we did not measure the real health outcomes of consuming UA produce but it is safe to say the regular provision of seasonal, plant-based food that is produced on clean soil, irrigated with clean water (see Chap. 6) and without synthetic fertilizers, pesticides and insecticides goes towards healthy eating habits. These results confirm previous research that points to the fact that gardening leads to diets of higher nutritional value and healthier eating behaviours, which extend to the gardeners' households and family members (Veen & Eiter, 2018; Tóth et al., 2018).

In a social inclusion context, it is important to raise the question as to whether all social groups can benefit from the opportunity UA offers to ensure healthier diets, especially in the light of the fact that socially significant diseases like diabetes and heart disease as well as obesity, are more common among the disadvantaged and poorer social groups, not least because unhealthy food that contributes to these diseases, such as high-fat, high-sugar and highly processed foods are disproportion-ately cheaper than their healthier alternatives (Garnett, 2000). There have been a number of studies that explore the chances for good life that UA can create in terms of helping individuals and households overcome certain life situations and habitual behaviours that put them at risk of poverty and social exclusion. Such situations include substance addictions and the burdens of a stressful lifestyle that homeless-ness, low education status, long-term unemployment and chronic illnesses can bring. There is indirect evidence that UA gives chances for a change of the social situation of disadvantaged individuals at risk of poverty and social inclusion by promoting self-esteem and life skills (Krasny & Doyle, 2002), offering therapeutic benefits through specially tailored rehabilitation (Garnett, 2000); stimulating mental health and bringing individuals and communities together (Prové, 2018). Still, even pre-suming those chances to be present, our goal was to see if UA in Sofia has shown real outcomes in helping any disadvantaged individuals overcome addictions, unem-ployment and other social deficits that put them at risk of exclusion and poverty.

So far, there has been only one organized attempt by an informal civil organiza-tion working as a charity, to involve disadvantaged people in UA in order to provide them with better chances of social transformation and social reintegration. The initial idea came from this charity which provides food for poor people in a solidarity kitchen twice a week at the same time as when the qualitative study within our research was taking place. An interview was taken from one of the initiators of the idea to establish what goals the organization had set in that respect and who is eligible for their help. In general, the work of the organization targets everyone who has financial and material difficulties and cannot live their life to their full potential:

Interviewee (R): they are so immersed in this daily life, to dumpster dive or, there are some without electricity and they live in quite difficult conditions. Without electricity and water, homeless, but even those who have a home but only live on a BGN 140[4] pension and all their money is spent on medicines have a difficult time. So that's why we started looking for initiatives that can both help them feel useful to society and accepted by society, and also to suppress their aggression because work, especially work with the soil, is quite calming and educational... this way they can help themselves,.... this is quite important as they have lost trust in others, they have lost belief and hope... and all these things that would help them improve their life(...)

Interviewer (I): What are the effects this could have on their lives, what do you think?... In practical terms?... How will their lives improve?

R: From a purely psychological point of view, it will first teach them to respect the earth more, and when they begin to respect the earth more, they will begin to respect all living beings more. Starting with plants, animals... this will make them more hard-working. When one works and does not idle around, then further progress starts. You don't just think about how to eat, because they are in this state at the moment, they only think about it, they only think about how they can find food, and when they are already full, this food also comes from their work... then they will find meaning in other things as well. And all in all, the thing is to plough the land, to grow vegetables, to see how your actions produce an effect... because they have lost hope and they think that nothing depends on them. It really isn't so, it all depends on them to sort out their lives, it does not depend on anyone else. So, in this way with urban agriculture, when they hoe their garden, plant their vegetables and see the result of what they have done, I think this will have a big effect on their mental state and their motivation to carry on and get out of that mess...

I: Have you shared this idea with the people themselves?

R: Yes, we have discussed it and we continue to repeat it as a mantra, because we believe that this is something positive. There are people who already want to go there, to see the garden.

I: So, there is interest?

R: Yes, and by March, when we will plant everything, we will inspire them, we will promote how good it is, how cool it is and of course, they only need to show an interest to see it. I believe, if only they see it once how it happens, they can't give it up. (from an interview with the coordinator of the informal charity)

At the time, the interviewee said that out of about 100 people who use their solidarity kitchen, there were three or four participants who had firmly said they wanted to participate in the gardening activities. Later in the year when the season started and

---

[4]BGN 140 which is equal to about EUR 70 and is completely insufficient to cover basic daily expenses in Bulgaria where the poverty line for 2018 when the interview was taken, was BGN 351. For 2018, 22% of Bulgaria's population lived under the poverty line.

in the following years not even one of the clients of the solidarity kitchen worked in the garden. In subsequent talks, the interviewee said he had not lost his hope integrating disadvantaged people through gardening could work, but he felt he had not done enough to motivate the people to get involved and had underestimated their more pressing daily needs to find food and shelter which takes all their time and efforts.[5]

Evident from this case is that even when there is good will in civil structures to create form of UA activity that helps integrate disadvantaged people, there is a need for more concerted efforts to explain to and motivate people to take up such activity that is new to them. It is important to underline here that in order to rely on UA for food security and through it – social integration, an individual or a household needs to have at least some buffer resources, such as time, tools and seeds, and some basic gardening skills. As many of the poor people are working poor (Ministry of Finances, 2018), they would have less time to spend on gardening. Also, gardening requires planning ahead of time and cannot be relied upon for urgent food needs. Therefore, any efforts to uplift disadvantaged people from poverty and social exclusion need to take these issues into account. These efforts could come from the local authorities or a stronger group of NGOs. The data from our research does not provide ground to support the claim that UA has the potential to improve significantly the quality of life of marginalized and socially excluded groups without deliberate top-down policies and local authority measures. This is because currently the disadvantaged social groups have either no easy access to UA activities, or do not have the time, skills, awareness or motivation to take part in them.

Still, the data from the representative study demonstrates that respondents from lower income categories are slightly more likely to engage in UA at present (see Fig. 5.3 in Chap. 5), which suggests that on a mass scale in Sofia UA is practised more commonly among the poorer social strata and therefore it could be suggested that it brings more economic benefits to them. However, those benefits are not related to enriched social capital and better social integration as long as these predominant practices are performed individually and do not lead to community building and social bonds. On the contrary, both from the in-depth interviews and the quantitative survey a conclusion might be drawn that just as it is difficult to integrate disadvantaged individuals through UA, urban agriculture activities have a social appeal to those who are already well integrated in society, with extensive social networks and high social capital. Those respondents from the interviews and focus groups who claim one of the benefits of UA are enriched social contacts and enlarged social networks, are in fact already well networked and participating in a wide range of other activities such as sports, environmental activism, etc. To the question on how one learns about UA sites and activities in Sofia and how one gets involved in them, an NGO representative shared the following:

---

[5] No interviews were organized with the users of the solidarity kitchen as methodologically it would have taken too much time to establish trust and rapport with them.

Table 4.1 Share of selected socially-related motivations to start UA according to the number of the respondent's acquaintances in the neighbourhood

Approximately how many people from the neighbourhood do you know by name?	Socially-related motivations to take up urban gardening		
	To share experience with others	To exchange seeds and varieties	To feel I belong to locals, neighbours
Up to 5 people	0	0.2%	0
Up to 10 people	0	0	0
More than 10 people	8.2%	3.3%	7.4%

> Well, where... it is in other campaigns for common good – afforestation campaigns, cleaning campaigns in the mountains, and... places like that, just word of mouth, the word is passed on and apparently people who are active in one area just start to be interested in another one, a third one and so on. At one point there is such a living chain in which people just share what they do and so on. (NGO representative)

The results from the quantitative study also point to the same conclusion – they reveal that the respondents with the highest level of social contact in their daily living environment (measured by how many people in their neighbourhood they know and how often they talk to people from their neighbourhood) represent the highest share of practitioners with social motivations to practice UA (Table 4.1).

This does not exclude the possibility for people to acquire social contacts and join social networks that improve their social capital once they have joined a collective UA activity, just as was demonstrated with the physical and mental health effects. However, while we identified a significant number of individuals who testify the positive, although unexpected effects for their health, the equivalent of a similar group of socially excluded people who have improved their social integration situation after taking up UA was not identified.

This raises the question as to what extent the commonly shared view that UA deepens social ties and increases social capital (Corcoran & Kettle, 2015; Prové, 2018) is realistic and to what extent the social capital identified among UA practitioners was a result of their UA activities and not in fact a pre-condition for them to take them up. One possible answer to this question is that this depends on how well developed UA is in the given location (city or region). Based on literature and our study in Sofia, it seems that in places where collective and socially significant practices of UA have a long history or they enjoy the active support of local authorities or civil groups, improving the social capital and social inclusion of disadvantaged groups can happen due to the fact that the UA networks have a much wider coverage, tradition and popularity and different disadvantaged groups are more aware of them or more often covered by civil society organizations who involve them in UA.

In Sofia, despite the very active awareness campaigns UA activists carry out, they are still few and UA, in all its forms, appears to be a concept largely unknown to citizens, as 41.4% of our respondents answered that they have absolutely no idea what UA stands for. When specifically asked about the potential positive effects of

UA, only 13% think of social benefits such as inclusion of disadvantaged groups, while the majority of respondents relate UA with producing healthy food (64%), educational benefits for children (43%), economic effects such as household budget savings (28%) and supporting relatives and friends with food products (18%).[6] Additionally, it must be pointed out that socially disadvantaged groups in Sofia include people who are on the brink of survival and without many social relations with any larger groups of people who share their fate.

This extreme social exclusion which portrays an aggregate of individuals who do not represent a community in the sociological sense of the word, makes it even more difficult to involve them in socially inclusive activities, as has been possible in other locations in Europe where deprivation and social deficits are shared within a social group, where gardening and shared diets function as a social bond (Veen & Eiter, 2018). Here, again, it is evident that if socially excluded individuals and households are to be encouraged to undertake UA activities that can lift them out of exclusion and poverty through enriching their social capital and through becoming engaged in an economic activity that contributes to their income, specially targeted policies are needed to compensate for these people's lack of capacity and skills which hinders them from seeing UA as a way out of deprivation and from making the best use of it in their interest.

As mentioned in Chap. 3, the satisfaction that individuals experience with their self-perceived capacity to overcome life challenges, regardless of their social, economic or health status, is a key indicator of personal quality of life (El Din, 2012; Tilkidzhiev, 2009). It accompanies a feeling of empowerment, which is particularly important for the elderly, as well as for young people and children, who are often thought of as more vulnerable groups than the middle aged. These effects on young people are analysed in more detail in the next Sect. 4.2. Here special attention is given to the external value of a good life which relates to how useful one's well-being is to others, as well as the environment. This is an especially relevant QoL dimension for the elderly who garden with younger people or who, through gardening, care for the healthy food their children and grandchildren consume. The elderly people who were interviewed or participated in the focus groups of our study testified that UA makes them feel better integrated with younger generations and feel their gardening knowledge and experience is valued by the larger UA community, which is seen as quite a young one. Having their advice accepted gladly by the less experienced younger gardeners and being recognized as the best gardeners (not least because they are retired, which allows them to spend a lot of time in their plots and maintain immaculate gardens), makes them feel important and appreciated. This, on its own, contributes to their feeling of living a useful life. This satisfaction and feeling of empowerment at an individual level is an important inner life result, directly linked to the belief in one's ability to exert control over and direct one's life, which is often lacking or decreasing when the elderly distance themselves from the labour market and the responsibilities of parenthood.

---

[6]The sum of the shares of different answers is over 100, as respondents gave more than one answer.

4.1.2   Community Level

The feeling of empowerment and civil participation at individual level has theoretically been linked to heightened levels of civic participation and lifting people out of poverty and social inclusion, hence, contributing to more equal societies (Tilkidzhiev, 2009). As already discussed, we do not argue against this theoretical proposition, but we report from our research that no evidence was found to prove this link: the urban gardeners who demonstrate active civil participation and empowerment have manifested these qualities before they took up UA. On the other hand, since gathering the empirical data in 2017–2019, some developments have taken place, including an increased membership of one of the community gardens we studied (ZaDruzhba) and the appearance of a completely new one (Community garden No 1). While we are not assuming that these transformed or new gardening groups can lead to social cohesion or improved neighbourhood community relations any more than the ones we had already studied, we have witnessed the formation of a small but tightly-knit community in the form of a pressure group that lobbies for UA regulations and targeted policies, supporting UA in Sofia. This group serves as a core around which social bonding and closer social relations are fostered, including with members of the garden who have not been active in the civic realm before.

As Veen and Eiter (2018) have demonstrated, even when social contacts are not a main motivation to start gardening, they do happen, just as social interaction happens between people who are often quite different in terms of occupation, age, value systems and outlook on life. Moreover, as gardeners often share produce, the exchange and reciprocity related to it would expectedly lead to enhanced social capital (Pourias et al., 2016). It is possible and likely that this happens, but in such a subtle manner that it could not be captured by our study. Due to the limitation of our research to only a snapshot of gardeners' social capital and only self-reporting on it, we cannot be sure that small and incremental contributions to neighbourhood communities or people's close social networks have not been made thanks to their UA activities. Our limitations are also linked to the very few examples of socially engaged urban gardeners: while 33.2% of our respondents believe that UA can bond people and 53.9% do not have an opinion on the topic, none of our respondents in the quantitative study actually practices UA in community gardens. Most of the UA practitioners (78.4%) use their own gardens and garden mainly with other members of the family or household (82.6%). The remaining 17.4% practise on their own.

This tendency of practising UA on private plots with immediate relatives paints a picture of an urban agriculture form that is strongly linked to bonding and not to bridging social capital. Among our respondents no one works in a community garden, but 1.9% of the participants in the survey know such people who collectively work in a vegetable garden in the green spaces between residential buildings. All this leads to the conclusion that when UA activities that could theoretically serve as a ground for enhanced social cohesion, inclusion and civil participation, are too sparse or with very little history, they have a weak potential to yield such socially significant results.

Apart from improving social cohesion through direct involvement in UA activities, we also tried to establish if cohesion and inclusion could be achieved through involvement of disadvantaged or socially excluded groups in activities that change the social profile of public spaces and through the mechanisms of the solidarity economy.

In the first case, people and groups that are commonly excluded from the daily life of a community, thus infringing their civil rights or hindering their participation in the decision-making processes that affects their QoL, are invited and allowed into public spaces where they can establish and get involved in social networks and become socially empowered. Just as we did not find evidence for social inclusion based on increased social contacts due to UA, we did not have substantial evidence of social inclusion built upon shared space-making. It could be hypothesized that place-making as well as social networking are reliant on some basic skills and social capital that, when missing, prevent reaching a positive effect of an activity like UA.

In the case of space-making, though, there is an interesting nuance in Sofia. While there are no local policies on making use of UA to integrate and support disadvantaged groups, there is some energy in the civil sector that works towards educating and socially integrating disadvantaged children in the safe space of their school yards, where vegetable gardens are built. Although the schools that have adopted such practices are still very few, they claim they are achieving some levels of social integration through UA. We have not followed the children who have gone through these programmes to measure the social integration effect on them. The coordinator of a programme that aims to integrate children with impaired hearing also does not claim they have become more socially included just because of the gardening activities, but this is their goal:

> Our goal is not only to teach the children how to produce food and what is the link with this production and conservation of nature, but also to make them somewhat more social, because living in this world of deafness, they are isolated from a large part of society. And this fact confines them to an isolated world from the cradle. We want to teach other people to communicate with these children, and the children to socialize with other people. (representative of an NGO that supports and develops vegetable gardens in schools)

The educational effects from these civil society actions are presented in more detail in the next Sect. 4.2.

In the case of the solidarity economy, where non-financial resources such as time, materials and specific skills, are used to provide food and services to people with little financial resources but with certain time and social skills, the case of Sofia has not so far been able to demonstrate successful results in terms of building communities around alternative economic, cultural and democratic opportunities.

A dimension of neighbourhood quality of life related to UA, but not based on personal involvement in UA activities, is the satisfaction with the neighbourhood. As demonstrated by El Din et al. (2012: 88), apart from the quality of social interactions, the levels of neighbourhood crime and public facilities, the quality of public green spaces is one of the best determinants of neighbourhood satisfaction. All of these could be impacted positively by UA. From our research we cannot deduce to what extent citizens' satisfaction with the green areas in their neighbourhood is impacted by UA activities, as we did not relate respondents' locations to UA locations. Still, from the in-depth study we see that people are very pleased with the landscapes that UA offers, which is always better than that of abandoned sites with rubbish piling up. More details about the spatial characteristics of UA and their influence on the QoL of citizens can be found in Chap. 7.

4.1.3 City Level

In order to have an impact on outcome dimensions of city QoL, UA needs to reach various communities, across different generations and areas of the city. At the city level, it also needs to be a subject of regulations or negotiations between different participants in city life: businesses, the civil sector and the local authorities. In the case of Sofia, UA has not achieved such a wide reach and momentum as to help it boost various city QoL elements reflecting measurable positive life results for the city dwellers. Still, there are a number of opportunities that exist for UA to unfold as a successful strategy towards increasing the QoL in Sofia, namely through a mobilized civil sector and through an improved governance of the city. This section of the chapter will analyse this two-way dynamics. The environmental and spatial dimensions of QoL at city level are discussed in Chaps. 6 and 7, while here we focus on two possibilities for UA to improve the QoL in the city. The first one is dependent on the goals and outcomes of the work of a range of NGOs which are active in developing UA initiatives directed to different social groups with the aim of educating them, improving their well-being and their chances for social inclusion and integration. The second one is related to the engagement of the local authorities with the strategic vision of UA as a strategy to improve citizens' QoL, and the specific local measures and steps they take to deliver results from this strategy. This viewpoint on the subject focuses on the larger social structures in the city that use or can use UA to achieve socially significant goals. In both cases we base our analysis on the presumption that although community-based and socially significant initiatives of UA are few and new in Sofia, individual and household practices of UA and gardening as a whole are a tradition in the country and in the city, and are very common for the citizens in the capital (Fig. 4.4).

Altogether, more than half of the respondents in the quantitative survey have practiced or are practising gardening at present (in the city, near it or in the countryside). Most of the practising gardeners have done so for more than 10 years. Keeping in mind that a little more than 40% of the urban gardeners and 48% of the gardeners who practise in the countryside do it to preserve a family tradition, we can safely assume that there is traditional gardening knowledge that is being preserved and passed down the generations, which is widely spread among the citizens of the capital and could be mobilized through civil initiatives, as well as centralized policies that can improve the QoL of the whole city.

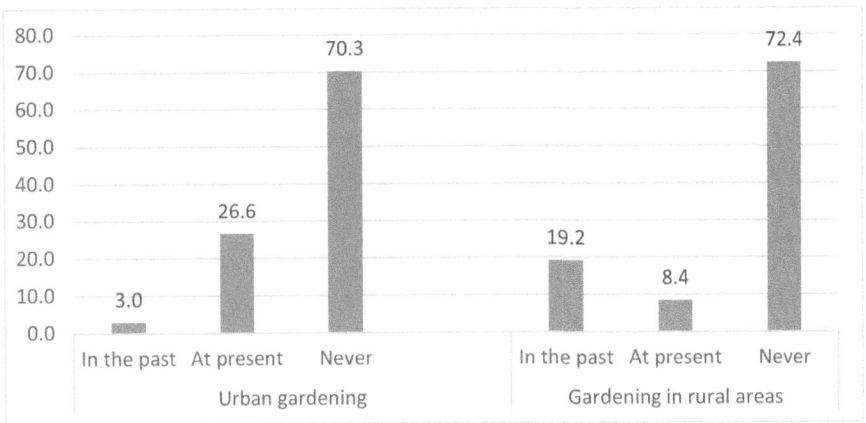

Fig. 4.4 Experience of gardening in urban and rural areas

4.1.4   The Potential of the Civil Sector and NGOs to Improve City QoL Through Agriculture

Within this project we have interviewed representatives of the four most active non-governmental organizations and initiatives involved in the practice and promotion of urban agriculture. One of these is not a formally registered legal entity, but still a very active and well-organized civil structure. This type of initiative develops on the principle 'from individual to collective and social': in the very beginning separate individuals realize the social value of urban agriculture, after which they share knowledge with other individuals of similar attitudes and gradually form small groups, sharing common values, attitudes and behavioural patterns. During in-depth interviews with representatives of NGOs, respondents speak openly about their desire to be socially useful by promoting urban agriculture and revealing the enormous socializing potential and the effects on healthy lifestyle, to help various social groups and society as a whole discover new opportunities to improve the quality of life through healthy and wholesome food, to develop new forms of social solidarity, by joining forces in making quality and healthy food, to show to people from different social groups that the creation of products is not only interesting and engaging, but it thus also brings about a healthy and useful output, builds social communities and relationships, educates individuals and expands the economic potential of the urban environment.

The most common goal of these NGOs is education aimed at raising awareness among children (and often their parents) about food, healthy eating habits and basic knowledge of growing vegetables. These effects are further discussed in Sect. 4.2. Another related goal is to link healthy food with the positive environmental impact of organic food growing and other methods of food production that are beneficial both for societies and for nature. The goal that we will concentrate on here is the socializing one – to develop vegetable urban gardens in order to stimulate people and communities to use food and the process of growing and consuming it to grow closer, to develop stronger social ties and social cohesion. As presented in the interview with a representative of one of the NGOs with the longest tradition of supporting UA which has worked in the field for more than 10 years, these relations are extremely strong and they grow into friendly ties which go beyond the initial common goals and spill over to other areas of social and economic life. Socializing effects to a large extent are related to the possibility of creating social networks or the formation of groups of people of shared needs who in the process of active interaction decide on how to meet those needs, by pooling ideas and available resources.

It is indisputable that the practice of urban agriculture can create lasting and sustainable models of social communication, build stable interpersonal relations over time and serve as the basis for the upgrade and the emergence of deeper social relations, affecting wider areas of the social and economic life of the modern Bulgarian. Still, as discussed earlier, no evidence was found that all citizens, irrespective of their social and economic status, have equal chances of getting involved in these networks, even when they have targeted specifically disadvantaged groups.

While very enthusiastic and devoted to UA development in the city, the NGOs themselves admit that their potential to drive real change in the QoL of citizens is too limited. This is not only due to the fact that their initiatives are few and far apart and their funding is not sufficient to address all people at risk of exclusion and poverty, but also because of the passive stance of the municipality, which can initiate socially-inclusive UA programmes on a much larger scale. Thus, a lot of opportunities for UA which Sofia has due to the plentiful resources suitable for UA, as described in Chap. 2, remain unused. One of the NGO respondents makes a guess that a third factor for the difficulties the NGOs face in trying to improve the overall QoL in Sofia, is that people generally prefer to take part in activities that bring quick financial returns and easy options rather than ones that require hard and persistent labour, such as UA. Even if there is funding earmarked for social inclusion and cohesion programmes, she believes these funds are mismanaged while at the same time the disadvantaged social groups and individuals remain living in desperate conditions, socially excluded and uninvolved in any socially significant activities:

> We're the epitome of the saying – scarcity amidst plenty. We are dying of thirst and starvation, in a figurative sense, but for some people this is literal – people in welfare centres, tramps, the homeless… Why doesn't the municipality – who else? – organize these people, these people should produce something, they should be given plots to grow something. The point is to be willing to do something. If you do not want to, it will never happen, you will only mark time, take money for nothing. We live in a country where decision-makers are unwilling to see things happen. They don't feel like making any effort, they are in for scams, interdependencies, string-pulling, subordination; in the end you get muddled and the result is naught. Good things happen despite the system, I'd say off state – all good things are happening off state. (a representative of a local food network)

4.1.5   *Local Governance*

External qualities of life chances relate to social, economic, environmental and other conditions for thriving, such as access to education, freedom, equality, social security, clean air and drinking water, which all have social outcomes. In Sofia, while the share of people living in poverty is lower than the average for the country, 20% living in material deprivation, compared to a 30% average for the country, inequalities are slightly higher, the Gini coefficient for Sofia being 42 compared to 40 for the country in general. Also, there is a territorial segregation where Western and Northern districts of the city display higher dissatisfaction with city life and are classified as more disadvantaged than the others (SofProject, 2019). In terms of life satisfaction, the citizens of Sofia are not completely satisfied[7] with a number of city QoL indicators that could be impacted by UA. These include public and green spaces, business opportunities, income levels, social services, nature and the quality of drinking water. Sofia citizens are even more critical[8] about the air quality, noise levels and the overall cleanliness of the city.

In the previous section of this chapter we demonstrated that these problems and inequalities have not so far been solved through UA but we have not refuted the possibility of this happening through targeted policies at local governance level. We gathered data on UA practitioners' evaluation of city governance, including that of UA, through in-depth interviews and focus groups. The results convincingly show that urban gardeners are not satisfied with the urban policies on land use and sustainable city development but they feel their gardening work contributes to the good life in the city. They believe urban gardening serves the public interest and the municipality "should also count the public interest and not only the private interests" (opinion from a focus group with urban gardeners). Considering the lack of any policy initiative on the side of the municipality, active urban gardeners organize themselves in formal or informal groups to have stronger leverage to negotiate with the municipality land and good conditions for using it for UA. This creates an important potential role of the civil sector to work towards social integration. However, without official policy strategy and regulation of these activities, there are also some risks that accompany such activities.

Business actors, and to a greater extent NGOs and community gardeners, carry out activities directly related to improving the quality of the urban environment (in social, environmental and economic terms). Some of them, especially NGOs, see their activities as overlapping with the goals of key strategic documents of local and state institutions (the Strategy for Sustainable Development of Sofia, the Strategy for Sustainable Development of Agriculture of the Ministry of Agriculture, etc.) and see the role of local government as a partner that could delegate responsibilities to those entities that have the expertise and capacity to support the work of the municipality

---

[7] The share of respondents who are completely satisfied with them is lower than the share of those with lower satisfaction.

[8] Total dissatisfaction outweighs partial and full satisfaction with these indicators.

with a view to achieving a number of strategic goals in social, economic and environmental aspects. Some respondents believe that there are enough enthusiasts and volunteers who already do the work of the municipality, or can do it, even without regulatory changes, but the municipality does not delegate these responsibilities to them, which again brings up the issue of the obstacles to civil society.

> They find experts to pay, they generate some proposals, but in the end, they don't put them into practice, and we are actually eager to implement them. They just have to figure out how to designate us to implement them without money – we go to them in person – we are passionate experts. Meanwhile they are looking for unbiased experts to tell them if this is good. /.../ The state and the municipality do not want to let go of the control and management, they want to do it, but in fact they do not realize the energy of the people – without paying salaries or funds, you can have people at your disposal who work. (representative of a food producer-consumer cooperative)

On the one hand, the local government recognizes the benefits of UA for a number of its policies (social, educational, environmental), but in reality it makes minimum efforts to maximize these benefits. On the other, within the study we have identified several examples of activities that have virtually identical goals to those set out in the municipality's policy documents, but are initiated and implemented by citizens or civic groups. Relevant examples are the 'solidarity kitchen' of the informal Food Not Bombs movement and an urban composting project aimed at developing social cohesion in small urban communities and limiting and recovering household waste. In both cases, initiators perceive them as activities to fill gaps in local policy, but these are often done in a completely unregulated manner and could even pose some health risks (e.g. cooking outdoors and without good sanitation for people in need). Interviews make it clear that these practices started in response to the lack of local policies for social support, education, environmental development, etc. In practice, strategic documents refer to such policy, but these actors do not perceive it as successful. As a result, there is a risk of some active civic groups taking over voluntarily the functions of local government (to integrate, improve quality of life – not only food and shelter, but also emotional recovery and mental strengthening, repair of roads, snow removal from local roads in winter), when in fact they do not have sufficient resources and competencies for this. While massive and full of enthusiasm, volunteer civil initiatives have limitations that can be overcome only by services that are highly professional and organized in a way that minimizes the risk of unwanted side effects (food safety, respect for the principle of non-harm in psychological assistance, etc.). In the absence of such services, however, entrepreneurial citizens resort to solving problems on their own, and this closes the vicious circle of "lack of activities and policies on the part of the local authorities – self-initiative of civic groups – deepening problems".

These perceptions among civic initiatives about local authorities distancing themselves from responsibilities, go hand in hand with the perception of local institutions as increasingly abdicating from their responsibilities to society. In response, civil society organizations either (1) take action seeking to work beyond the influence of the local government, to go beyond the scope of their regulations and policies, to "get rid" of them, or (2) are actively fighting to change the *status quo* –

i.e. with policy instruments for policy change – for example, the Civic initiative for urban agriculture led by the Gorichka Foundation which actively lobbies before the Municipal Council for regulation of urban agriculture as a legitimate activity to improve the urban environment and quality of life of urban communities. At the moment, no real result of these activities can be reported, but given that the pressure on the municipality has intensified only in the past year, we cannot rule out the potential of the civil movement for UA to contribute to qualitative changes in the city's governance model. What exactly do these practising actors expect from the municipality?

As for citizens, they see local government in various actual and potential roles. In general, the prevailing opinion is that the municipality does not do enough to promote UA, and should do much more, instead. Interesting evidence of this opinion of citizens can be found in the Expert Assessment provided for our study by a representative of the local government, a representative of the Sofia Development Association, in which he states the commitment of the municipality to UA development, but relevant examples refer only to the conversations in which he participates with the civil sector, which sought help for UA development initiatives, which at the time of writing this analysis are carried out without the support of the municipality. The need for a clear commitment to UA through the regulation of related activities and expectations for policies, rather than financial resources, is shared by a large number of civil actors.

> The (farmers') market in front of the Ministry of Agriculture (. . .) has been supported by the municipality with money. We have never wanted money and most likely will not want any. We think there should be cooperation – when you have it, resources are made available. Once we have 50 people around – we can raise, for example, BGN 10,000 from producers tomorrow! The municipality has given them this amount. I.e. money is not what we need most, although it would be better to have it, but rather policies, a change in some working conditions. . . (representative of a food producer-consumer cooperative)

It should be noted that according to most business representatives, the main role of municipalities and the central government (the Ministry of Agriculture, Food and Forestry) would be to promote UA among students/children – in educational institutions, as well as to allot plots for urban agriculture, designated for this activity, including the development of a business model for UA. Several UA practicing businesses also believe that the role of the authorities is to inform the public about the importance of healthy eating and the risks of using foods harmful to health. The rest, they believe, should be their job – it is a kind of self-fulfilment.

On the side of the local authorities, it is evident from the interviews and focus groups with their representatives that there is a very positive attitude to UA, but these attitudes are expressed only as a personal opinion and the respondents are convinced that the municipal institutions do not have to do anything more than they are already doing in the frame of the status quo of municipality policies, strategies and daily responsibilities. This position creates a serious cleavage between UA activists and municipality officials and policy-makers and without much constructive dialogue between these two types of public actors, UA does not stand a good chance of developing in a way that is beneficial for the whole city.

4.2   Educational Effects of Urban Agriculture

The educational effects of urban agriculture prove to be among its most highly valued and widely accepted contributions to improving the quality of life of urban communities according to the participants in our study. All categories of respondents in the project surveys identify the multifaceted educational impact as one of the unconditional benefits of urban farming.

4.2.1   Educational Institutions as the Main Channel for Transfer of Educational Effects of Urban Agriculture on an Individual Level

The results of the research show that educational institutions are among the pivotal structures in society which enable agriculture in an urban environment and more and more modern projects related to the development of urban agriculture are implemented under their umbrella. Given their main functions and purpose, it becomes clear why the educational dimensions and effects of urban agriculture fully unfold through these institutions. The educational institutions in the lower levels of education – preschool, primary, basic and secondary education – are most commonly involved in UA practices in Sofia Municipality. A meeting between the research team and the mayor and representatives of the municipal administration of Mladost Municipality (one of the three districts where a representative survey polled the population) showed that urban gardens, mainly with vegetables and herbs have been established in almost all kindergartens across the municipality. Meanwhile, higher education institutions are to a lesser extent involved in urban agriculture; the two case studies of university-level UA practices[9] are among the rare cases of existing practices of urban farming in higher education in the capital city.

The analysis of the educational effects of urban agriculture here is based on the results of in-depth interviews and focus group discussions conducted within the project. Out of a total of 25 in-depth interviews, four were with representatives of educational institutions. Two interviews were conducted with representatives of the gardening project of Prof. Dr. Decho Denev Secondary Special School for Children with Hearing Impairment, where urban farming practices are part of the educational process, one with representatives of the University of Forestry, Sofia, more specifically the Training and experimental field station in Vrazhdebna district on the outskirts of Sofia, and one representative of Sofia University's Botanical garden in Sofia city centre. An interview was also conducted with a representative of the ZAEDNO

---

[9]The training and experimental field station of the University of Forestry – Sofia and the University Botanical Garden Sofia at Sofia University "St Kliment Ohridski".

foundation, which has set up a school garden at the 107th Khan Krum Primary School. Out of a total of five focus group discussions, three were with representatives of educational institutions – one of the focus groups was entirely with staff members of educational institutions – and the other two were mixed, with the participation of educators from secondary schools and higher education institutions.

When it comes to educational institutions as places engaged in urban agriculture, quite logically their main goal and sought effect is educational. The *educational effect in this case targets the youngest residents in Sofia Municipality – children, pupils and students.* However, the in-depth interviews and focus groups conducted show that there are still differences in the type of urban farming practices and respectively in the type of resultant and dominant effects depending on the type of educational institution. Thus, for schools and kindergartens, the most typical practice of urban agriculture is plant gardening – growing mainly vegetables and herbs, sometimes fruit trees, for educational purposes and not for the production and satisfaction of food needs, while higher education institutions are characterized by urban farming practices, which include, in addition to plant growing the breeding of animals, mainly for educational purposes, but also for production and sale of agricultural output and income generation.

4.2.2 Educational Effects of School Gardening

The educational effects of UA are especially visible in *school gardening* as a practice of urban agriculture, typical of primary, basic and secondary education institutions, where due to the complete absence of market motivation to practise urban farming, its main functions and targets are purely relevant to the educational process. All participants in the focus group discussions and interviews insist that the main *raison d'etre* of school gardening is not production (the nutritional effect for students) or achieving economic profit for schools. This form of UA is not aimed at meeting economic needs. The main function of school gardening is educational; it is intended to educate children and pass on knowledge about plants and nature, the path of food from sowing to table, the care that plants need to grow and bear fruit. Thus, involvement in urban farming activities teaches students about how food is produced, and they can apply this knowledge in other settings (e.g. at home, or to opt for professional farming in future), thus multiplying the educational effects of the practice of urban agriculture in schools.

Involving students in food production, making them part of this process, allows them to get acquainted with the true path of food, restores the connection with the earth and nature, which they have lost amid the modern lifestyle, where you can buy everything in stores.

I want my child to see where food comes from. When I ask her: where food comes from, I don't want to hear – from the store. And when I ask her: what cows are like – to get the

answer: purple.[10] When my own daughter asked me, "Mom, which tree bears tomatoes?",
I realized that the child was not to blame, she had seen tomato (pictures) in the kindergarten
glued on the board, and had no idea of the process whereby the tomato get to her plate on the
table. (a representative of the Zaedno Foundation)

Therefore, the knowledge about seeds, plants, fruits, vegetables, animals, which
children gain at school, remains purely theoretical for them, textbook-acquired
knowledge, without any practical experience, and therefore not relevant in their
practice and daily life. Urban agriculture, school gardens, take children away from
computer screens, tablets and phones where they spend most of their time, including
as part of the learning process, and gives them a new type of productive task and thus
real knowledge through experience. This is therefore sustainable knowledge and can
then be applied in practice in situations outside school.

> Children learn about nature by looking at slides and textbooks, and I have had cases of
> children claiming that cheese is a plant-based product, and they aren't referring to the latest
> trends in cheese production.[11] These are urban kids who, especially if they have no relatives
> in the countryside, don't have a clue. (manager of the educational programmes of the
> botanical gardens of Sofia University)

Losing touch with the land, with the village, which happens to most urban children
nowadays, leads to the loss of hands-on-knowledge, the knowledge of how to plant
and take care of the plant so that it grows and bears fruit. And if there are no places to
practise this skill, there is a risk of completely losing it. School gardens give children
a chance to experience the whole process of plant growing – from sowing the seed in
the ground, through growing it, providing it with clean soil, water and sun, to
harvesting fruits and collecting seeds to resume the whole process the following
year. Hence, the practice of UA at school becomes a never-ending lesson from
nature. In this sense, school gardening serves as integrated education where students
go through the whole process of growing the plant, including knowing which other
plants it can coexist with, looking for lessons at every step, and bonding with people
and society. For example, just as plants hinder or support each other, so relationships
between people are built on similar principles. In this way, mastering the lessons of
nature becomes a way to learn the lessons of society, to understand the lives of
people in society, and this repeatedly multiplies the educational effects of urban
agriculture, which gain new dimensions.

At the same time, providing opportunities to put in place the principles of learning
through experience, school gardens are the place where children's education
strongly resembles alternative education methods popular worldwide, such as

---

[10] One of the most popular milk chocolate brands in Bulgaria has a purple cow as its symbol.

[11] This refers to the widespread practice in the past more than 20 years of dairy products with
vegetable fats being sold on the Bulgarian market in order to reduce production cost and achieve
more competitive prices, which, however, affects the quality of products, misleads consumers and
in some cases has negative consequences for their health. In the past 2 or 3 years, the Bulgarian state
has relatively successfully opposed this practice by forcing producers to offer such products on
special stands, separate from dairy products, to label them with special tags and impose severe
sanctions for non-compliance with these requirements.

Waldorf's and Montessori's, whose specificity is expressed in learning and creating real knowledge through experience. Thus, the school garden, even when it does not specialize in the application of such alternative methodology, becomes a practical field, creating opportunities to acquire knowledge of various sciences, such as mathematics, biology, languages (languages are taught in the garden, mathematics through counting seeds, petals, etc.), thus reaffirming its function to contribute to the integrated education of pupils.

However, the educational effects of urban agriculture in schools, learning about the true path of food and the ways of its production are not limited to knowledge acquisition alone. A very important positive effect, as identified by the surveys of the project, is the fact that acquired knowledge becomes a prerequisite for the transition from the level of knowledge to the level of practical behaviour and respectively to the formation of new models of practical behaviour in children and students, promoting more ecologically oriented and responsible behaviour.

> There lives a boy. . . in my neighbourhood. . . I started teaching him to plant tomatoes and he got interested. . . He plants one tomato, one cucumber. Interestingly, the plants can be manipulated. For example, if you don't water a flowering tomato too much, it's like telling it, "A bad year is coming." And in order to survive, it begins to bloom more. When a tomato is 'pampered', it bears larger fruit, but fewer; in poor conditions it bears more, but smaller ones. I told that to the child. For him, it's like a game. And he says, "Let's make this tomato bigger, and this one smaller." The child does not think of it as a job at all, he does not perceive it as a job, he is interested in it, because children are interested in everything. (a participant in the focus group discussions on urban agriculture)

So, besides teaching children environmentally-oriented behaviour, work in school gardens teaches them to take responsibility for their actions, calls for their proactive behaviour, teaches them to be active individuals able to look for and make decisions. Thus, the educational effects of urban agriculture go well beyond the narrow educational functions, in the sense of transmission and acquisition of specific knowledge and affects the overall development and formation of children as future active citizens in society, developing a range of skills closely related to the inner dimensions of QoL chances.

> They learn where food comes from, what will happen when hail strikes, they learn to accept crisis situations as something normal, they learn to look for solutions, not to give up. (a representative of the Zaedno Foundation)

But changes in some patterns of behaviour in turn lead to changes in other patterns of conduct resulting from or related to the former. In this case, the formation of models of proactive and environmentally responsible behaviour in students results in a change in consumption patterns, expressed in the interruption of the consumer model of behaviour imposed by retail chains, which encourages you to buy everything from the store; this is replaced by new, more responsible consumption patterns, based on the awareness of the connection between food and the earth and that in order to have food in the store, it must first be produced.

These new behavior patterns fall within the outer QoL results as such. In this sense, it can be argued that we have a complete educational cycle, from acquisition of knowledge and skills to the formation of values and implementation of relevant patterns of behaviour, which means that urban agriculture as practised in educational institutions, fulfils its purpose and leads to positive educational effects – both as life chances and as life results. While the existence of such positive educational effects is proclaimed by the representatives of the institutions and organizations we included in the research, we have not measured these effects ourselves by involving pupils and students in the study for ethical reasons. Still, 43% of the respondents – a significant share of the public in the three districts we included in our quantitative survey – state that one of the most important benefits of urban agriculture is "children's acquisition of knowledge and skills in how to grow plants". When specifically prompted to share their opinion if they approve of the idea of schoolyards and kindergartens providing a plot of land for children to grow edible plants, 79.9% of the respondents express agreement with it.

Another aspect of the educational effects of urban agriculture in schools, in addition to the acquisition of knowledge, is the formation of specific social skills in students, the first one being the skill to work.

> Now this happens in many countries – in Germany, in Japan, in Switzerland there are school gardens. Developed societies such as Germany and Japan know that for society to prosper, children must learn to work – this teaches them work and discipline themselves. (a teacher who took part in a focus group discussion on urban agriculture)

Urban agriculture in the school garden teaches the work ethic. From the cradle a child is an object of care, but in order to become a fully-fledged citizen of society, they must first learn to take care of themselves and food production is one of these ways. And by learning to take care of themselves, they become able to take care of others, making the transition from individual to social utility. Thus, although economic benefits are not the main intended effect of school gardening, there comes a time when the educational benefits grow into economic benefits, and the educational effects are multiplied.

The most important social skills that urban agriculture forms in children include: skills for joint activities, teamwork, assistance. These are all basic, but especially important social skills, which express the ability, attitude and readiness for social cooperation, as a particularly important prerequisite for successful future inclusion and life in society.

> Children who work together while doing activities in the school garden naturally communicate more with each other – exchange experiences, knowledge. In my opinion, this helps enhance trust among them. Children learn to work together and rely on each other. (the director of the school for hearing-impaired children)
>
> Working in this garden, children make contacts, rely on mutual help – between younger and older… They help each other… and so they feel happier, more connected, involved in shared activities. (an educator at the school for hearing-impaired children)

The statements of participants in the focus group discussions and interviews clearly show that the joint work of students in school gardens shapes their skills for social cooperation and assistance not only in respect to their classmates with whom they work together. Indeed, this creates sustainable patterns of behaviour, which students then apply in respect of other vulnerable groups and communities to which they express a readiness to help, providing them with products of their labour. Thus, school gardening develops in children a sense of social responsibility and empathy for the needs of others and evokes a sense of belonging to a larger social community.

> I think there are countless lessons to learn. Initially, children asked me: if we grow a kilo of tomatoes, who would eat them; and I told them that if there were so many of us and the yield was so little, we would give them to an elderly lady. Thus, their thoughts opened up to new horizons which some of the children had not even thought of so far – that you may be happy by making someone else happy. And the children are very proud to have grown something they can share with someone else. (a representative of Zaedno Foundation)

Thus, the outcome of school gardening strengthens students' trust in each other and their tolerance for the others, all these being qualities testifying to the formation of social connection, so necessary for the prosperity of society. The multiplier effect of the educational benefits of urban agriculture once again illustrates its ability to contribute to improving the quality of life of urban communities and society as a whole, creating and increasing social cohesion and connectivity.

And, related to the formation of specific social skills in children under the influence of urban agriculture at school, we cannot overlook the fact that urban agriculture helps to build a new type of skills for students to communicate with their teachers and parents based on shared interest and joint activities in urban agriculture.

> I observe them in our school, the children are those to make their parents get involved. That's how we got to know each other – both teachers and parents – we see each other digging, doing different chores, that's how we got to know each other, that's how we made friends. (a teacher who participated in a focus group on urban agriculture)

> We created a community by doing several campaigns in which parents, teachers and children made this garden together. (a representative of the Zaedno Foundation)

Thus, school gardening turns out to be the weld that brings together children, teachers and parents, puts the relationship between them on another level and allows them to overcome or at least significantly limit some of the problems in the parent-teacher relationship today, when both are increasingly involved and therefore try to take responsibility for children's upbringing. School gardens show that the solution is in mutual support and joint action, in creating a parent-teacher community, because both belong to the same society, which is responsible for their children. Probably that is why the majority of respondents in the quantitative survey support the application of urban farming practices in schools and kindergartens. As mentioned, nearly 80% of all respondents believe that it is good to have areas where children can grow plants in kindergartens and schools and only 3.3% see this as unnecessary. All this speaks of a high level of awareness in society of the educational benefits of urban agriculture.

4.2.3   Educational Effects of Urban Agriculture on University Students

The primary role of the educational effects of urban agriculture practised within educational institutions is confirmed by the representatives of higher education institutions interviewed in the survey – the Training and Experimental Field Station at the University of Forestry (TEFS at UF) and the Botanical Garden at the Sofia University (BG at SU). And since in this case the educational impacts target mainly university students who, unlike pupils, are already developed personalities, the educational effects here are manifested mostly in the narrow educational sphere, namely in the acquisition of new knowledge, in the implementation of research, in carrying out experiments and changing production technologies in the direction of more environmentally friendly production.

> In fact, educational activity is our priority, because this is a training and experimental field station. Students have practical seminars here. University professors also carry out research here. Many projects are being developed on the basis of the experimental field station. Doctoral dissertations are developed on the basis of experiments and research work in the experimental field station. (the director of TEFS of UF)

Highlighting the educational effects of urban agriculture, representatives of higher education institutions also note its advantages over mere acquisition of knowledge from textbooks, slides, educational films. These advantages are related to the overall perception and contact with nature, which can be achieved only by direct involvement in activities and practices of urban agriculture.

> Students need to be able to experience what they study. When we talk about the educational function, it affects all senses, not only the knowledge about the plant, there is also aroma, there is touch, there is emotion. (a representative of the Botanical Garden – Sofia)

At the same time, the fact that the educational effects of urban agriculture in higher education are aimed primarily at the acquisition of knowledge, does not exclude its effects on the formation of students' specific social skills, namely the formation of communication skills, networking, cooperation and mutual assistance, which lead to the achievement of new levels of social cohesion and trust among students, which in fact once again outlines the exceptional integrative role of urban agriculture and its capacity to increase and multiply social benefits for society.

> Working in the field together, students gradually learn to help each other, begin to plan who will do what in the future, who will help whom... They start thinking about how to start a business and who will do what, so that together they can help each other in the future. (a lecturer at the TEFS at UF)

At the same time, joint activities involving urban agriculture contribute to the formation of students' communication skills not only among themselves but also between them and their lecturers, giving them a new sense of belonging to the community, which is built not only on a professional basis and the teacher-student relationship, but also on the basis of their joint activities with agriculture.

It is an established fact, moreover, that the practice of urban agriculture within higher education institutions has an educational effect not only with regard to direct participants in the educational process in higher education, i.e. students. For example, as school groups have the opportunity to visit the training and experimental field station of the University of Forestry, namely for cognitive purposes, the educational benefits of this form of urban agriculture are extended to a wider range of users, building a bridge, a lasting connection between different types of educational institutions – primary/secondary schools and higher education institutions: this creates a sense of unity/community of the educational system and its participants.

This aspect of the educational effects of urban agriculture is also emphasized by the representatives of the Botanical Garden, Sofia, who are working on the project *"Demonstration hives: bees in the city"*, namely going beyond the narrow confines of the educational institution and transferring educational effects to a wider circle of users, expanding the functions of education achieved through urban agriculture to education for the general public.

> Since in the past, botanical gardens were much more closed and were used only for scientific purposes, we have tried to expand their function and at the proposal of the World Council of Botanical Gardens to task them with a social function – education for the general public... We work more on providing knowledge and information. (a representative of the BG at SU)

Working on the project *"Demonstration hives: bees in the city"*, the representatives of the Botanical Garden, Sofia are trying to acquaint the general public with the path of food and the methods of food production.

Thus, the analysis of the interviews and focus group discussions clearly shows that the potential educational effects of agriculture in an urban environment are indisputable, irrespective of the type of the educational institution where it is practised and the relevant field of education – primary, secondary or higher, while these effects almost always go beyond the specific institution and the direct participants in the educational process and have an impact on a wider range of social actors, thus extending the educational benefits of urban agriculture.

4.2.4 Educational Effects of Urban Agriculture on the General Public

The analysis of the educational effects of urban agriculture so far has outlined the most important path of its impact – on children and youth, which expresses the greatest added value it has for society – the upbringing of the new generation, new members of the future society, sharing new understandings, values, attitudes and patterns of behaviour. However, urban agriculture has an educational effect not only for the youngest – children, pupils, students, it has a similar impact on adults and older members of society. This is confirmed by the statements of teachers in schools involved in urban agriculture, who share the view that work in school gardening is a source of new knowledge not only for students but also for themselves. In order to be able to teach students, they themselves must first be educated in the field of urban agriculture, because for most of them this is a new field of interest, which they do not know well enough yet. And in the process of practicing urban agriculture, they gradually learn more and more new things and acquire more and more new knowledge, including from their students.

The process of education in urban agriculture is therefore becoming on the one hand, mutual exchange of knowledge between teachers and students, on the other, relationships among teachers themselves, engaged in urban agriculture. The exchange of information between them begins to build not only on the basis of their professional knowledge, but also on the basis of their knowledge of urban agriculture. The parents of pupils are gradually involved in this process through the so-called culinary festivals, for example, which almost all schools with gardens organize when the time to harvest comes. These culinary holidays perform an educational and cognitive function, not only for students, but also for their parents insofar as they allow them to get acquainted with the revival of old traditions, to try different traditional recipes to prepare products made by their children. All this in a very definite way confirms the educational effects of urban agriculture not only with regard to the youngest, but also in relation to the mature and already socialized and formed members of society.

Although the educational institutions illustrate in the most complex way the educational effects of urban agriculture, we can talk about such effects in other practices of urban agriculture applied in Sofia Municipality. These are illustrated by the effects of practising agriculture in an urban environment in another form – shared and communal gardens, created in the inter-block spaces or on other vacant areas. These gardens become a place to pass on knowledge and experience between generations, affirming the value of each of the generations and the importance of the specific experience it brings to the practice of urban agriculture.

> Grannies got involved in Vasko's garden and helped him a lot in the first years with a lot of love. They came to him to watch how he worked and were ready to give advice… But for Vladi grannies proved invaluable, because they came every day to see if he planted plants properly and ask why not use ammonium nitrate for tomatoes. And Vasko started teaching the elderly ladies and explaining why ammonium nitrate is not advisable as a fertilizer, which was a taxing and difficult job. But all this did not bring any negative emotions, on the contrary. (a participant in a communal garden)

One of the biggest advantages of urban agriculture is that it helps restore knowledge and skills relevant to food production among young people living in cities, while at the same time it is a perfect way to share knowledge between young and old. On the one hand, there is a transfer of knowledge, skills and experience from older to younger – older people, having lived in another time, were engaged in agriculture, they know how to do it, while young people who have never been engaged in agriculture need to learn, and someone should show them how to do it. On the other hand, it is a process of knowledge transfer the other way round – from the younger, those familiar with new views, concepts, with new ways of food production, to the older. Thus, urban agriculture emerges as a continuous process of learning and exchange of knowledge and skills between different generations and between people with different social status.

> A lot of people came in the first two years, and a lot of elderly ladies came to argue with us –
> why raised beds? – they had their own ideas about gardening, but they treated us and our
> ideas with respect. To date, they take care of the garden – they water, weed, do chores.
> (an initiator and creator of a community garden)

And in a broader sense, the educational effect of urban agriculture on the quality of life of urban communities (in the case of the inhabitants of the capital city) is related to the impact it has on building specific social skills for communication and interaction between people from different backgrounds, ages and generations in the process of its practice.

> At first it was harder for me to communicate with the elderly, but here in the garden I have to
> and gradually got used to it. They show a desire to help you for free, to teach you something.
> I learned to take their advice more easily. I see the elderly in a new light as they are more
> experienced than me. In my opinion, this is very important for younger people, because we
> are very arrogant. (a participant in a communal garden).

This wider educational effect is associated not only with the acquisition of inter-generation communication skills, but also with the acquisition of such important social skills as cooperation, mutual assistance, the ability to trust and rely on the other, to accept help from them, to understand that one is not alone. Similar to the educational effects of UA in schools, where students learn the norms and rules of coexistence and communication, cooperation and mutual assistance in addition to their work in school gardening, work in communal gardens also teaches people who are already grown-ups cooperation and mutual help, it forms in them the feeling that "you are never alone", that there is someone to rely on.

> When it's dry, you water yours your (plants), you water (the plants of) the others... we rely
> on each other. No one is alone in the garden. (...) When someone is absent, when someone
> is sick, we try to take care of their land. We are a community, we rely on each other, we trust
> each other. (a participant in a communal garden)

At the same time, statements of participants in such gardens show that mastering these important social skills goes far beyond the narrow framework of cooperation, mutual assistance and trust in the process of working in the garden and is transferred to the wider social life of people involved in urban agriculture. Mastering the skills for joint activity in the process of food production, for cooperative actions, also forms skills for trust, for communication, for building social connections and contacts in the broadest sense with people of different ages and professions and with different social status. Thus, on top of curiosity about food, it brings about curiosity about people and interaction with them, and in addition to studying food and the ways of its production, people also get to know each other.

Some of the respondents in the surveys overtly point out that the main purpose for which the communal gardens are created is precisely the educational one, and educational effects target not only adults, but also children, as far as they, being still developing personalities, are more susceptible to educational influences, while the knowledge they acquire becomes a model of behaviour in the long run, unlike adults, who as already developed personalities, are more unsusceptible to change and educational impacts.

> We put a stronger emphasis on the educational effect of our activities. . . we decided that it is much better for children to have non-formal education; they have classes at school, so they can have classes in this garden to learn where food comes from . . . It is best to bet on children as a development for the future – what they remember now, they will remember in 10 or 20 years, it will stay in them. . . Sometimes it is difficult to work with people over 35, because they are resilient personalities, it is very difficult to inspire them, while with a child this is the moment and we therefore focus on children. (an initiator and creator of a communal garden)

Thus, urban agriculture for the people involved in it turns from mere food production to the process of cooperation, mutual assistance, building relationships and networking among people of different generations and with different social status, mastering the principles of social relations and interaction and affirmation of the social principle in humans. By involving both elderly and young people, communal gardens in spaces between residential buildings become both a way to pass on knowledge and experience between generations, and a means of bonding. They turn out to be a unifier of people of all ages – from children, the middle generation, to the elderly, thus creating a community of people of different ages, each of whom, however, derives specific benefits for themselves from their participation in this garden – one learns, another one passes on their experience and thus a community is formed where everyone has a certain function and everyone is useful with something for others.

> Urban agriculture is a kind of people's development. Communication between people, being good at something, makes you less aggressive. And you probably will not break the neighbour's windshield, if he parked on your spot in front of the apartment building, as some time ago you watered tomatoes together. (a participant in a focus group discussion on urban agriculture)

Here's how joint food production becomes a way of educating people in the basic principles of social coexistence in the way of activation of social communication, investment in the social development of people.

4.3   Urban Agriculture and Sustainable Consumption Patterns

This section of examines the relationship between urban agriculture (UA) as a social and cultural phenomenon and practices, and the formation of models for responsible and sustainable consumption in a highly urbanized environment, such as the country's capital. In this regard, we discovered and collected empirical evidence how through changes in the value motivation and orientation of citizens, in the transformation from conventional, unsustainable to sustainable consumption, models of healthy eating and respect for food, a sensitive attitude and behavior to nature, natural resources and the environment are formed, environmental models that ultimately reflect on the QoL of urban communities.

4.3.1 Sustainable Consumption and Quality of Life

The last few decades have seen a boom in economic, technological, informational progress and growth of the world economy, accompanied by a significant improvement in the living conditions and lifestyle of the humankind. On the other hand, there is a global trend towards an increase in the urban population, and hence the needs and increasingly tempting human desires to diversify consumption, especially material consumption. Internationally, all this has led to overuse or depletion of natural resources, to climate change and ultimately to environmental degradation. It became obvious that one of the key elements for achieving sustainable development is the transition towards sustainable consumption and production in a way to produce goods and services more efficiently with lower risks to humankind and the environment (UNEP, 2015a: 3). Thus the notion and transition of sustainable consumption embodies the relationship between business activities, consumer behavior, and environmental and social challenges (World Business Council for Sustainable Development, 2011: 1).

Sustainable consumption is permanently in the focus of the international policy organizations. It was first accepted as a goal at the 1992 Rio Earth Summit, (NGO Committee on Education Agenda, 1992), further developed at the Johannesburg *World Summit on Sustainable Development 2002, and has since been present in the* UN's 17 Sustainable Development Goals (SDGs) aiming to improve the planet and the lives of its citizens by 2030 in Goal 12: Sustainable consumption and production patterns. Its targets seeks to achieve sustainable management and efficient use of natural resources; to ensure that people everywhere have the relevant information and awareness for sustainable development and lifestyles in harmony with nature (UN, 2015). Defined in the Oslo Symposium (Ofstad et al., 1994) it means "the use of goods and services that respond to basic needs and bring a better quality of life, while minimizing the use of natural resources, toxic materials and emissions of waste and pollutants over the life cycle, so as not to jeopardize the needs of future generations".

Among the various definitions of sustainable consumption, the shortest one is "doing more and better with less" (UNEP, 2021). It means not to consume less but to consume better, i.e. more efficiently, with less risk to our health and environment, to consume responsibly (UNEP, 2015b). Sustainable consumption has multiple dimensions – environmental, social and economic ones (Šajn, 2020; OECD, 2008). Sustainable consumption can generate economic benefits, social wellbeing and social inclusion, such as access to markets, innovation, job creation, healthier livelihoods and lifestyles (UNEP, 2015b), and also protect the environment and natural resources available long into the future (Srinivas, 2015). To ensure the sustainability both of the production and consumption, their life-cycle stages must be considered (UNEP, 2015a). Sustainable consumption is not simply an act of consumption of goods and services. It is mindful consumption (Lim, 2017), a complex set of mental and behavioral characteristics of individuals' and group attitudes, values, orientations, way of thinking, etc. Sustainable consumption is also a choice to prefer such goods in which not the price is the priority, but their quality although sometimes they do not look "shiny".

Orientation towards cleaner or "green" production and sustainable food is a linchpin of EU strategy and policy to achieving the SDGs (SDSN and IEEP, 2019). It is based on the goals to boost the demand for sustainable products, but also to help consumers make informed choices' and to tackle the challenges related to "climate, biodiversity, and circular economy, as well as in strengthening the convergence in living standards, across countries and regions" (SDSN and IEEP, 2019).

To understand sustainable consumption, it is important to trace the link between consumer values and sustainable consumer behaviour. The transition from conventional to sustainable consumption is directly related to lifestyle and consumption, but is no less determined by subjective characteristics of the individual and social group such as values, motivation, conscious behaviour and more.

Sustainable consumer behaviour or green purchase behavior is central to building sustainable consumption. The UNEP has defined sustainable consumer behavior (SCB) according to the human's main life 'functions' as nutrition, housing (e.g. sustainable building, energy and water conservation, etc.), clothing, education, healthy and environment-friendly lifestyles and leisure (UNEP, 2002, cited in Kostadinova, 2016). White et al. broadly consider sustainable consumer behaviour as "actions that result in decreases in adverse environmental impacts as well as decreased utilization of natural resources across the lifecycle of the product, behavior, or service" (2019: 24). Remi Trudel also stresses SCB is a behaviour that attempts to satisfy present needs while simultaneously benefiting or limiting environmental impact (2019).

Sustainable consumer behaviour is a process that is determined by factors of different origin and scope. In the literature, a set of psychological factors are considered as the basis of a shift framework to sustainable consumer behavior: social influence, habits, individual self, feelings and cognition, and tangibility. Social factors are one of the most influential motivational drivers for people – among them stand out such as social identities (group membership), social norms and social desirability, social environment, demographic characteristics, etc. Another set of drivers are the cognition-related ones, such as information, learning, and knowledge. Habits are also a very strong factor for a change of behaviour and where they dominate, the change in behaviour towards sustainable consumption would be slower, as in general habits are unsustainable but are more enduring (White et al., 2019: 25–28; Joshi & Rahman, 2015; Figueroa-García et al., 2018). Most often the drivers are a combination of different factors – social, psychological, ethical. Green purchase behaviour factors can be analysed at different levels, such as individual – like emotions, habits, values and personal norms, trust knowledge, and situational factors such as price, product availability, subjective norms, social norms and reference groups, product attributes and quality, eco labelling and certification (Joshi & Rahman, 2015: 132–134; Šajn, 2020).

Consumer values and environment-friendly attitudes are seen as major predictor of sustainable consumption and behavior with a view to outlining the organic connection and process of transformation to the implementation of SDG 12 for sustainable life style (Vergragt et al., 2016). Values inherent or embodied in the individuals and groups are the basis in the course of building consumer behavioural norms, but they are the same ones that can change them through the mind-set of consumers' from conservation values (tradition, security and conformity) to open-ness to change values (Martin & Upham, 2016: 208). In addition, individuals' internalized ethical rules about what is right or wrong can also influence the consumption choices they make (Šajn, 2020: 3).

Referring to the conceptual model and definitions of ten basic values according to their motivational goals, developed by Shalom Schwartz (1992), we used three of them as relevant values that guide individuals or communities of people who are convinced and ready for transition to sustainable consumption or who already practise it. More specifically, these are self-direction (creativity, freedom, choosing one's own goals, independent thought and action, etc.); stimulation (choosing, creating, exploring excitement, novelty, and challenge in life) and hedonism as a "pleasure and sensuous gratification for oneself" (Schwartz, 1992: 6–8).

This institutional and theoretical framework directly relates to our topic of sustainable consumption patterns in urban agriculture practices. As stated in the previous sections of the book, one of the functions of the UA is to inform, educate and form new values, attitudes, and motivations for turning to sustainable behaviour and consumption. In this sense, UA consumer patterns are strongly linked to the implementation of SDG 12 and its sub-objectives.

In the context of our research subject we consider sustainable consumption as consumption of products and food produced in a way that is gentle on nature and natural resources; conscious and internalized (by individuals or groups) consump-tion, motivated by values and oriented towards sustainability in its broadest sense (environment, technology, lifestyle, health, etc.)

4.3.2 Sustainable Consumption, UA and Quality of Life

By this definition we assume that consumption of UA products and food that are produced by traditional, non-industrial, safe and soil- and land-friendly technologies, in the most natural and nature-friendly way, foster sustainable consumption patterns.

We agree that sustainable consumption "minimizes environment effect (. . .) and is for the satisfaction of needs that produce a better QoL" (Kilbourne et al., 1997: 5) and present the ways in which sustainable consumption of UA products affects QoL at the three levels presented in Chap. 3.

At the individual level it is a consumption of good quality, clean and healthy food; better health and spiritual status; increased awareness, values, behavior, new knowledge and experience.

Here we do not analyse sustainable consumption at the community level, because we did not collect enough reliable information that could allow us to shape commu-nities' sustainable consumption patterns.

Table 4.2 Interlinkage urban agriculture, sustainable consumption and QoL

	QoL – Individual level	QoL – City level impact
Sustainable consumption of UA products/food	Consumption of quality, clean healthy food; cultivating respect for food; better health and spiritual status; awareness, values, sustainable behaviour, knowledge and skills	Better environment; minimization or lack of harmful effects on soil, water, air, etc.; greening neighbourhoods; sustainable consumption networks

At the city level, indirectly this type of consumption is expressed in nature protection, non – or less – waste of natural resources for production and minimization of harmful effects on soil, water, air, etc., achieving a cleaner living environment (neighbourhood, urban spaces).

In this section of the book the impact of the sustainable consumption on the dimensions of QoL is analysed mostly at the individual level and at the city level, only insofar as it is present in the opinions of the interviewees (Table 4.2).

4.3.3   Data and Analysis

In this section we consider only certain aspects of sustainable consumption such as attitudes towards production and consumption, the quality of products and food consumed; attitudes towards natural resources and the environment; attitudes and transformation towards sustainable behavior. We will illustrate some real-life examples of sustainable consumption identified in our fieldwork. We will look for an answer to how UA can form sustainable consumption patterns, and who the main actors of this consumption are.

In this text, the models of sustainable consumption are constructed indirectly, mainly through indirect evidence, collected through the in-depth interviews that we have conducted with various stakeholders – key representatives of producers and consumers of products from UA, organizers of farmers' markets and food cooperatives, civic groups and NGOs engaged in UA activities.

We did not research specific sustainable consumers target groups but some of our respondents are such consumers themselves. The empirical data collected represent the summarized opinions, assessments and ideas of the respondents with different profiles. These consumption patterns can also be derived from producer-consumer relationships, for which there is empirical evidence.

The objects of sustainable consumption in our study were UA products and activities and the actors in this consumption – the people practising UA in the urban and suburban areas of the City of Sofia. The products of UA under study are from the communal gardens for UA, the individual and family farms in the yards, the small agricultural farms. The sustainable consumers and actors mediating the sustainable production and consumption patterns are food distributors, farmers' markets organizers and chefs.

4.3.4 Patterns of Sustainable Consumption

The models of sustainable consumption and main actors of sustainable consumption products, according to the results of the study, can be summarized in several types of sustainable practices.

4.3.4.1 Producing and Consuming One's Own or Family's Production

This is the most common model of sustainable consumption in the context of the topic we studied. Here are representatives of individual or family producers, hobby farmers and producers from the urban and suburban areas of the city who sell part of their yield on the market. These are people who grow their own products in an environmentally friendly way for their own needs and thus consume sustainably. Most of them have their own experience or experience in agriculture from their families. They are influenced/driven by external and social factors, traditions, habits, etc. Such an example is a producer for his own use in a communal garden in the city. He produces for the needs of his family. He believes in the value of his self-produced food.

> They [the products] are not for sale or distribution. Variety and cleanliness are valued. We eat them, share them with our loved ones. We have a small baby. We have given the products to friends who feed their children – in various forums, groups. There is no economic dimension, the goal is to teach people where bread comes from, as the saying goes, to grow it themselves and have fun, because this is creativity. (gardener producing for himself in a communal garden)

Another small producer of ecological vegetables by extensive mode without any chemicals was also motivated to start farming "by the search for clean food".

> My family has always grown food in our own gardens, but it has not been plentiful, and I have gone to many farmers' markets and shops in search of a variety of food; I have seen that prices are very high and this is not affordable for everyone. That's why we got into it, we saw that it was in demand. . . . I am excited, I am interested in healthy eating; I see that most young people are excited about healthy eating and looking for it. But not at such high prices as in shops. (a small, young producer of vegetables for her family and the market)

A new group of mostly highly educated and well-informed people, most of them well-paid professionals who have no experience in agriculture but have embraced the idea of producing themselves and thus to consume sustainably, can also be referred to this type of producers. Factors linked to the individual self (White et al., 2019: 27) have a stronger influence on the formation of their sustainable consumption behaviour. This group is also more driven by cognition.

> Currently more and more people have realized that they want to somewhat go back to nature and a natural lifestyle, for them fitness or fun in unsustainable consumption or passive leisure in front of TV is not enough, but they need rather something to do with their hands, something that comes easiest in nature – to plant something and then grow it. . . . These are very young people. . . – first, they are aware and know to some extent what they need to do, i.e. they are beyond the stage of initial learning, e.g. we have done workshops in the

farmers' markets for making seedlings.... most people do not know anything about the
stages of food production. (chairman of a cooperative, organizer of farmers' markets)

These are people who value freedom, creativity. It is important to have a simple life, to
produce our own food, or at least to know how to do it. All this gives us meaning as people.
These people are a bit alternative, freelancers, they are into extreme sports. (gardener
producing for himself in a communal garden)

The results of the sociological survey confirm that individual or family producers
constitute the largest recognizable group of consumers of products from UA. In
response to the question about the main consumers of urban agricultural products,
the most common answer was "people who produce them", given in 81.6% of all
cases, followed by "relatives, friends and acquaintances of producers", indicated in
42.4% of all responses. Only 4% of all responses point out that other people who
consider these products cleaner consume them. People knowledgeable about the
production of urban agriculture output were named as main users in only 2.7% of all
responses.

With regard to the motivation behind urban agriculture (Fig. 4.1) predominant
motives of the respondents were strongly linked to sustainable consumption. The
desire to consume healthy products was shared by 64.7% of respondents, the
pleasure of producing food themselves by 66.2% and confidence in the high taste
quality of home-made products by 49.3% of respondents. A strong motivation also
related to the pursuit of healthy eating, referring in particular to the concern about
children's health in the family (34.8%). An important group of motives is related to
the pursuit of personal physical and mental health – 15.9% and 21.9% – to ease
mental and emotional tension.

4.3.4.2   Farmers' Markets as Distributing Channels and Space
              for Sustainable Consumption

The way to sustainable consumption is the growing potential of urban agriculture for
bringing producers and consumers closer together, providing customers with a
selection of vegetables in-season from local farms (Deelstra & Girardet, 2000). In
our case this most often happens through farmers' markets which, in recent years,
have grown and are increasingly becoming a sustainable production and consumption model of local, biologically pure or organic products produced by certified
companies or producers of food of verified, proven origin. Such markets have
established themselves as centres where the high quality of the offered produce is
truly guaranteed. Their popularity is growing, as well as the number of regular
clients. The representative of one of the municipal companies responsible for
organizing the city markets claimed that at farmers' markets in Sofia, 100% of the
suppliers are small producers, "with 0.5 hectares of land and 5 cows", and as many as
90% of them are from suburban farms, which makes the control of their production
processes easier.

> ... farmers' markets (...) are also a form of sustainable consumption. They provide access to a range of goods in a certain place for a certain short period time... The farmers' market gives us a guarantee that the market organizer knows that this is an organic product, i.e. when I go there, I have no doubt what I will find. I won't ask myself if this cheese is organic or not, I know I'm going for this type of product.... and I don't go there only to buy tomatoes, but I check also celery or fresh butter or something else. (representative of the local government, a sustainable consumer)

The formation of sustainable consumption patterns among consumers is strengthened in the producers-consumers' communication after building trust between them. The experience and ability of the producers of high-quality products to promote them to consumers makes the latter more aware and convinced of the value of good food.

This group of sustainable consumers often has more interest in the food-growing process and the biophysical processes involved when crops are locally cultivated. This could increase the influence citizens have over the way food is produced. People will understand what sort of inputs are used in the farming process and they can quickly respond to any harmful environmental practices. Thus urban agriculture can re-educate about the ecological base of food, and the links of food production to natural food chains, as well (Deelstra & Girardet, 2000: 54).

This is how our respondent from the cooperative, organizer of several farmers' markets in the City of Sofia, commented on the way of establishing sustainable production and building sustainable consumption. He said that farmers' markets shorten the producer-consumer relationship (in knowledge, communication, trust). The consumer can learn directly from the producer how a product is made; if they cannot explain it, they cannot sell it:

> ...we record some conversations between producers and consumers. This is part of our selection – if they don't explain how they do it [produce an item] and why they do it – they won't be able to sell it. There are such cases – they come and keep silent, don't know the exact variety and they don't sell. In order to be able to sell, you need to know and explain. I call them educated farmers. Our farmers are like talking machines – they already have experience and constantly explain, give recipes, explain what they have seen, why exactly, when they did it, how to store it for longer. (chairman of a cooperative, organizer of farmers' markets)

> Very quickly the producer gets information about what (produce to do), how to do it and how to price it, i.e. they get feedback – they launch the product and in the first week they can find out if it will sell well. Out of more than 50 producers, there are probably more than 40 examples of people who started with one product and then marketed 15 that they chose (to produce) and most of them are unique. Especially for the vegetables market, 2 years ago, this was a very weak sector – they made the lowest revenues, and now I have the feeling that they are stronger than dairy. Apparently, people have learned and have confidence that their quality is good and as early as 9 o'clock there are queues for vegetables, while before there used to be queues for dairy products. (chairman of a cooperative, organizer of farmers' markets)

4.3.4.3 The Profile of the Direct Consumers at Farmers' Markets

Studies in the field of consumption behaviours show that most of the individuals likely to be engaged in pro-environmental behavior in their consumer choices are younger, more liberal, and highly educated individuals (White et al., 2019: 28).

This trend is also confirmed by our study from the interviews with producers who offer "green" products (through direct deliveries, short chains, farmers' markets or otherwise), as well as from those interviews conducted with other stakeholders in UA. The profile of these customers can be identified as "green purchasers" of products and foods produced in a sustainable way. Most interviewees describe these target customers as young people, families with small children, aware citizens with a responsible attitude to food and healthy eating.

This profile is confirmed by a young female farmer, a regular seller at one of the farmers' markets who already has a network of regular customers, some of whom directly order the desired products from her by phone in advance. On the other hand, this case is a vivid example of established lasting communication and trust between producer and consumer:

> Personally we rely a lot on regular customers we have met over this year and two or three months. We have about ten regular customers, some call in advance, others are first to come on the market, they come especially to us... They call us regularly, they ask whether we have been hit by hail, for example; they are happy if we have a new product. And there are more and more people of this kind. It is a great pleasure for them to know that they buy from a person who has grown it with love. Shopping is a nice emotion amid the grey urban life.

She already recognizes the individual tastes and preferences, figuring out the profile (social and age status) of these customers and tries to meet their needs with healthy products.

> We have regular customers, people who are friendly, in a good mood, but mostly these are laid back people who can afford such food – it is a bit pricey, still not too expensive. But anyway, the person who comes to us has a clear idea of what organic farming stands for. Mothers also come to buy quality products for babies as well as for their families. People are between 20 and 45–50 years old... They appear to be educated, we always have a conversation with them. They don't just come to buy and leave, but they are regulars – they come back, there is hardly anyone who has failed to return... The elderly are rather few, maybe because they do not have such information..., but retirees can hardly afford it. People over 50 perhaps represent 15% of all our customers. (young, small-scale producer of vegetables for own consumption and for sale)

Only a small part of consumers ask where the production comes from. For most it is important that it is not treated and fertilized with harmful substances.

According to a micro-greens producer, consumers of clean products are mostly more educated households of "university degree holders who are young, have a child and want to give clean vegetables to their child, especially those with small babies". For example, he quoted the observations of a colleague – a certified organic producer:

> Every spring and summer there is a boom of young mothers looking for carrots, zucchini, potatoes, vegetables good for infant purees… And every year there are new customers because there are always mothers anxious about chemicals in vegetables in the big store and looking for an organic producer. … . . so everyone – from the concerned mother to a normal family conscious of what they eat. (producer of micro and baby vegetables)

A representative of the municipal company "Markets", which claims to support UA practices on the territory of the city, partners of farmers' market organizers and other NGOs with an environmental focus, shared her observations from the latest organized food festivals in Sofia – according to her thousands of young people could be seen there for whom clean food has already become a way of life.

A producer of animal products on a suburban farm profiled sustainable consumers in an interesting statement. He sells his output to a certain circle of people who contact him through social networks or through close acquaintances and friends. His main customers are "intelligent people, above average intelligence, who are well-off, people of broad horizons and rich life experience who opt for clean food". According to him, consumers choose his products mainly because the area is environmentally clean, pastures are mostly in forests and products are of high quality. By contrast, he claims that "ordinary folks" are not into this type of food:

> Well, it's more convenient for them to go to the store, buy everything and eat soy, palm oil, starch and milk powder and transglucominosis, and they don't care at all. Otherwise, one has to come to me, wait, call me, because I can't do deliveries – there is simply no one to do it. (producer of livestock products on a suburban farm)

According to another respondent, head of an NGO promoting organic agriculture and producing organic output, raising awareness among consumers as to what healthy food and a healthier lifestyle are like and only information about the shelf life of food is not enough:

> … unless they see for themselves or make the difference between organic production and products, vegetables and foods that they see in supermarkets… and unless they know how these are produced, they could never understand why they get ill after that.

She does not trust non-certified production if she does not know the producer in person.

> I want to see how he proceeds. So I always ask questions of a new producer at the farmers' market. I ask such hard questions that I understand if they do their job well.

4.3.4.4   Public Restaurants as an Environment for Sustainable Consumption

Here we sum up data from in-depth interviews with representatives of restaurants as "collective consumers" of urban agricultural products from the region of Sofia, as well as foods of organic origin, which they offer themselves. One of the restaurants we studied is not labelled as a gourmet restaurant, but rather as a "funky bistro" "in the European sense of the word".

The way it looks, the things we prepare and the way we treat our guests show that it is not a "prime time", but a funky restaurant, where traditional Bulgarian recipes are prepared in a very modern way, and food has a certain terroir. (restaurant manager, representative of public consumer of urban agricultural products)

In this type of restaurant, food is used as a tool to educate and form sustainable taste habits and attitudes towards food produced from natural products from non-industrial production, which is the basis for building sustainable consumer behaviour – not only in terms of respect for food and its preparation, but also in a broader sense of behaviour for environmental protection, for the sparing use of natural resources. Observations in the studied restaurant show that irrespective of their social status, guests have in common enhanced interest, a desire for awareness of what is offered to them in the restaurant, where it is produced, how the food is prepared, which speaks of a change in the culture of consumption of the city dweller. In other words, there is increased pressure from consumers to producers and intermediaries in the preparation of food in public places. This requires enhanced responsibility, trust and (self-)control of each actor in the chain, encouraging competition between them to win over customers.

People are already quite open and pay attention to what they eat. We first give them information about what we serve as a dish and products, but their interest is already very keen... People want to be more aware and sometimes even ask profound questions about the technique of processing the product, what is the breed of meat, what is the ratio of cheese (percentage of cow or sheep), percentage of fat. (a restaurant manager)

In another restaurant of this type, the owner and chef is guided by the belief that people who come to his restaurant to feel good should always be served with personal care and attitude, which is why the customers in the restaurant are also guests. According to him, these are people who consciously come to eat at his restaurant because "they know exactly where they are coming and what to expect". In this way sustainable consumption attitudes and the reciprocal customer services enforce each other and create a fruitful space for sustainable production and consumption patterns.

A restaurant from the Slow Food network, while not a typical example of a place that largely uses urban agricultural products, is very similar in design, purpose, spirit and operation to the other restaurants we studied. It is based on the idea of building a restaurant that would develop "the connection between people seeking sincerity and purity in food". The restaurant adheres to "sustainable practices such as shortening the food's path to the plate, minimizing waste and stimulating local production", principles embodied in the concept of the mission of urban agriculture and the sustainable consumption practices.

In all three restaurants mentioned, the use of fresh and seasonal products is a priority and a label of quality and guarantees origin and variety in the menu of the food offered. In the Slow Food restaurant, seasonal cuisine and hand-made products are prevalent and come from small and family associations and farms that grow them in a natural way.

The modern way of interpreting food reinforces the uniqueness of otherwise "boring and ordinary" products that can easily be produced in an urban environment. Modern and environmentally friendly technologies of processing and presentation of products in such restaurants is another step towards a new attitude towards food, forming sustainable patterns of production and production.

4.3.4.5   Online Purchases Through Internet Platforms

In Bulgaria there are many online platforms for food and products of different range, origin and scope. Here we focus on a specific Internet platform "Home delivery of clean food. Genuine products from small farms and producers", which in addition to delivery, aims to make the producer-consumer relationship, in which customer feedback (opinion, evaluation) is very important in achieving the goal of the platform: delivery of quality goods and food of clear origin. The platform guarantees the origin of the foods that are offered because of their certificates or the inspection performed. This builds trust between producers, distributors and consumer and is a prerequisite for sustainability of this type of consumption. The principles of co-operation, saving of transport and other resources, that ultimately impact on the protection of the environment, are also observed here.

Likewise, as at the farmers' markets the platform works in a similar way. There, the pre-selection done by the organizers informs the customer which goods are from certified companies or companies whose certification is underway. Thus, the consumer can both visit the platform deliberately for a product they usually buy, but also if they see a new item that attracts their attention, the trust the website has built in the customer will be transferred on to the product and better the chances that they buy it:

> I log in to look for dairy products and the platform offers me various groups of products... but when you enter you find something and see an additional set of goods that can attract your interest and these are all foods about which I am certain because of their certificates, the inspection performed. (sustainable consumer using an online platform, representative of the local government)

Taste, quality and healthiness of products have been reported as important attributes for consumers who purchase green food products (Cerjak et al., 2010, cited by Joshi & Rahman, 2015).

Transition to sustainable consumption is a process, a trend back to 'consuming to live' from of 'living to consume' (by Kilbourne et al., 1997).

Orientation towards better and healthier food consumption is growing and gaining supporters, although this share of consumers is still too small, there are no aggregate social data for it to be analysed. The main bearers of sustainable consumption in all its forms which we have studied are mainly (but not only) the more educated, young people and families, with a value orientation towards healthier nutrition, higher requirements for origin and qualities of the food consumed. According to the survey data, in the scale of respondents' values, "consumption of quality food" (ecologically clean, with an established producer or of local origin) ranks third (48% of all responses) in the ranking of values in terms of importance among all respondents, after family safety and financial security, being valued as very important by 63.6% and 59.5% respectively.

Through sustainable consumption urban agriculture has the potential and role for value transformation – transition from material (economic) benefits and habits (traditional livelihood, employment, self-sufficiency, income, etc.) to post-material values (environmental awareness and behaviour, togetherness, orientation to a healthier lifestyle (physically and mentally) There is a value transformation in the individuals' attitude towards food and its origin; changes in purchasing habits – the demand for clean food increases and forms patterns of responsible consumption. In general, the sustainable consumption of UA products and foods reinforces the values of sustainable development and higher quality of life.

Sustainable consumption implies not only purchasing behaviour, but includes all types of interactions between individuals and infrastructures (mobility, leisure, housing), which together make up lifestyles and livelihoods (UNEP, 2015b). According to the survey data, the motives for mobility (out of the city) among the respondents are more related to sustainable consumption of a clean environment (the desire for peace and privacy and the desire to communicate with nature). A total of 24.6% of respondents who would like to leave the city and move to the countryside are divided between the desire to grow fruit and vegetables (9.1%), to raise their children in a healthy environment (7.9%) and to prepare ecological food themselves (7.6%).

It should be borne in mind that eco-labeled products are in a higher price range and not every consumer, value-oriented to this type of consumption could afford them. Thus, the assertion that "income level is not a criterion for embracing more responsible consumption" (Nevison, 2008: 23) is not valid for the great majority of the Bulgarians who simply could not afford those products even if they aspire to the sustainability values epitomized by them. This situational factor (price) combined with social and demographic factors are more valid as a motive for the representatives of those sustainable consumers with experience in agriculture. For them, the price is one of the drivers – to produce high quality food might be cheaper than buying it, but still to consume clean and healthy food is a higher priority. On the contrary, for the group of young and educated people the price marker is not a leading factor for sustainable consumption, or it is absent altogether.

In less affluent countries like Bulgaria, individual consumer purchasing is very price sensitive. It is unrealistic to expect an individual consumer to set aside price considerations in favour of "loftier" principles of social justice and environmental protection when he or she may be struggling to meet basic survival needs (UNEP, 2015a).

Therefore, it can be summarized that the sustainable consumption of UA products taking the example of Sofia City, in the various models discussed here, is still in its infancy stage, but we remain convinced that the values and motivation for healthier eating and better quality of life, regardless of how they can be achieved, will become a priority for wider social groups of the population with few or minimum resources.

4.4 Lessons Learned

The analysis of the effect of UA on social cohesion, social inclusion, developing inner and outer QoL chances and results for young people through UA education activities, as well as its potential role in fostering sustainable consumption behaviour, was done by looking into the building elements of these QoL dimensions.

We found a link between urban gardening and physical health (including improved diets for the producers and their relatives and close friends) and a particularly strong relation between UA and improved mental and psychological well-being. Both of these QoL dimensions contribute to the chances of individuals to be better integrated in the social and economic systems of the city, and enjoy being active participants in the community life and being empowered to control their own life path.

Urban agriculture activities were also found to be central for the perceived personal capacities to overcome life challenges for more vulnerable social groups such as children and the elderly, as well as contribute to their enhanced satisfaction with life. Including structured UA activities in the daily schedule of schoolchildren and kindergarten pupils seems to have a role in the formation of personal qualities and skills, such as an improved work ethic, heightened responsibility towards their peers, raised environmental awareness and increased tendencies to follow sustainable consumption patterns – all inner QoL chances for a better life. The data collected on the inter-generational social connections that are established in the studied community gardens unambiguously point to a feeling of better integration and recognition experienced by the elderly gardeners who, in a generally young gardening community, feel valued and important for their knowledge and experience of gardening and life in general.

On the other hand, when there are deep inequalities in an urban society, and individuals who are entirely outside the safety net of a social network, UA activities on their own are both very challenging to perform because of time and resource scarcity, and very unlikely to lead to social integration and improved overall social cohesion in a community and in society as a whole. Such an effect could be expected when targeted and wide-reaching policies support awareness raising, initial motivation and first steps in UA for disadvantaged groups. Otherwise, the current profile of communal UA practices, most commonly represented by young and middle-aged, educated and well-integrated in society, is likely to remain the same, without making use of the opportunities to involve disadvantaged social groups that either have no easy access to UA activities, or do not have the awareness or motivation to take part in them. The potential to integrate these groups and better their life conditions through UA seems difficult without focused top-down policies, which are completely missing at the moment.

Our findings on the effect of UA on social capital reveal that in the cases similar to Sofia, where gardening activities are performed mainly in a tight family or household context and rarely in communal gardens where interactions are broader, the type of social capital affected is not bridging SC, which brings individuals and communities from different backgrounds and resources together, but it relies on and enriches bonding SC within families and narrow social circles of relatives and friends.

As to place-making and improving the QoL dimensions related to the inclusiveness and social value of urban space, our findings reflect the importance of providing safe space for various social groups to practise UA in a supportive environment that puts every participant on an equal stand. This could be a responsibility of a public authority or institutions in charge of regulating the use of a particular space, public green spaces and school yards. Similarly, in order for UA practices to develop their potential to improve the QoL of individuals with little financial resources through the mechanisms of an alternative economy, built upon relations of exchange of non-monetary goods and assets, initial support from the local authorities is needed to regulate these exchanges, as well as a well-connected network of coordinators who make this alternative system run smoothly. In all cases, the balance between the role of the civil sector and local authorities in directing the evolutionary path of UA in a city or its districts was demonstrated to be of great importance. When there is no constructive dialogue between all types of UA stakeholders, the synergies needed to take UA from a practice without social significance to a factor for improving the QoL of urban communities cannot be established.

The research findings lead to the conclusion that when UA activities that could theoretically serve as a ground for enhanced social cohesion, inclusion and civil participation, are too sparse or with a very short history, they have a very weak potential to yield socially significant results. Still, there are specific enabling and disabling factors identified in the framework of the study, which can contribute to UA's potential to improve urban QoL, or to stifle it:

- Participation and engagement of all stakeholders – citizens, municipality and the non-governmental sectoral, is seen as key to sustainable UA initiatives. On the one hand, top-down policies are not expected to work without grassroots engagement, but on the other, bottom-up initiatives face too many uncertainties if they are not supported by official policies. The non-governmental sector is seen as a necessary mediator providing know-how and network connections for producers without easy access to markets and deep knowledge of agro-ecological methods of production.

- There is a marked need for political support for recognizing UA and providing a binding normative framework that regulates the use of public spaces for such practices. If UA is to provide more encompassing benefits to the urban communities (including the disadvantaged groups), this cannot happen without the vision and long-term strategies of the municipality and the public institutions. So far, political support by the municipality has been based on impulsive enthusiasm and 'wishful thinking' and not on binding rules and regulations. In this way there is no security for gardens, farmers' markets, etc. While the activists insist on rules and specific local legislation, the municipality officials claim there is no need for such policies and their declared support is enough. A lot of our respondents believe that the political adoption of such new ideas is always slow and it is only natural that it will take time and a lot of effort to achieve it.

References

Cerjak, M., Mesić, Ž., Kopić, M., Kovačić, D., & Markovina, J. (2010). What motivates consumers to buy organic food: Comparison of Croatia, Bosnia Herzegovina, and Slovenia. *Journal of Food Products Marketing, 16*(3), 278–292. https://doi.org/10.1080/10454446.2010.484745

Corcoran, M., & Kettle, P. (2015). Urban agriculture, civil interfaces and moving beyond difference: The experiences of plot holders in Dublin and Belfast. *Local Environment, 20*(10), 1215–1230. https://doi.org/10.1080/13549839.2015.1038228

Deelstra, T., & Girardet, H. (2000). *Urban agriculture and sustainable cities*. Thematic Paper 2, MILUnet, Multifunctional Intensive Land Use network. Retrieved on June 21, 2021, from https://www.researchgate.net/publication/284992045_Urban_agriculture_and_sustainable_cities

Draper, C., & Freedman, D. (2010). Review and analysis of the benefits, purposes, and motivations associated with community gardening in the United States. *Journal of Community Practice, 18*(4), 458–492. https://doi.org/10.1080/10705422.2010.519682

El Din, H. S., Shalaby, A., Farouh, H. E., & Elariane, S. A. (2012). Principles of urban quality of life for a neighbourhood. *Housing and Building National Research Center Journal, 9*(1), 86–92. https://doi.org/10.1016/j.hbrcj.2013.02.007

Figueroa-García, E. C., García-Machado, J. J., & Pérez-Bustamante Yábar, D. C. (2018). Modeling the social factors that determine sustainable consumption behavior in the community of Madrid. *Sustainability, 10*(8), 1–16.

Garnett, T. (2000). Urban agriculture in London: Rethinking our food economy. In M. Bakker, S. Gündel, U. Sabel-Koschella, & H. de Zeeuw (Eds.), *Growing cities, growing food: Urban agriculture on the policy agenda* (pp. 477–501). German Foundation for International Development.

Grootaert, C. D., Narayan, V., Jones, N., & Woolcock, M. (Eds.). (2004). *Measuring social capital. Integrated Questionnaire World Bank.*

Joshi, Y., & Rahman, Z. (2015). Factors affecting green purchase behaviour and future research directions. *International Strategic Management Review, 3*(1–2), 128–143. https://doi.org/10.1016/j.ism.2015.04.001

Kilbourne, W., Mcdonagh, P., & Protero, A. (1997). Sustainable consumption and the quality of life: A macromarketing challenge to the dominant social paradigm. *Journal of Macromarketing, 17*(1), 4–24. https://doi.org/10.1177/027614679701700103

Kostadinova, E. (2016). Sustainable consumer behavior: Literature overview. *Economic Alternatives, Issue, 2*, 224–234.

Krasny, M., & Doyle, R. (2002). Participatory approaches to program development and engaging youth in research: The case of an inter-generational urban community gardening program. *Journal of Extension, 40*(5). Retrieved June 12, 2021, from https://archives.joe.org/joe/2002october/a3.php

Lim, W. M. (2017). Inside the sustainable consumption theoretical toolbox: Critical concepts for sustainability, consumption, and marketing. *Journal of Business Research, 78*, 69–80.

Martin, C., & Upham, P. (2016). Grassroots social innovation and the mobilization of values in collaborative consumption: A conceptual model. *Journal of Cleaner Production, 134*(A), 204–213. https://doi.org/10.1016/j.jclepro.2015.04.062

Ministry of Finances of Bulgaria. (2018). *Written answer by the Minister of Finances to a parliamentary question on the income level of the employed persons according to groups and categories for the year 2017*. Outgoing letter No 01-00-249 from 10 October 2018. Retrieved June 12, 2021, from https://www.parliament.bg/pub/PK/318729854-06-1098.pdf (in Bulgarian)

Nevison, J. H. (2008). *Impacts of sustainable consumption choices on quality of life: The Slow Food example*. Master's degree thesis, University of Columbia. Retrieved on June 21, 2021, from 56373079.pdf (core.ac.uk)

NGO Committee on Education. (1992). *Agenda 21, Chapter 4*. UN Documents Cooperation Circles. Retrieved on 28 May, 2021, from http://www.un-documents.net/a21-04.htm

OECD. (2008). *Promoting sustainable consumption. Good practices in OECD countries.* Retrieved on June 21, 2021, from 001-002-999-16x23eng.fm (oecd.org)

Ofstad, S., Westly, L., & Bratelli, T. (1994). *Symposium: Sustainable consumption: 19-20 January 1994 : Oslo, Norway.* Ministry of Environment of Norway.

Pickard, D. (2013). *Social capital as a factor for development of rural communities in Bulgaria.* PhD Dissertation. Prof. Marin Drinov Academic Publisher.

Pourias, J., Aubry, C., & Duchemin, E. (2016). Is food a motivation for urban gardeners? Multifunctionality and the relative importance of the food function in urban collective gardens of Paris and Montreal. *Agriculture and Human Values, 33*(2), 257–273. https://doi.org/10.1007/s10460-015-9606-y

Prové, C. (2018). *The politics of urban agriculture: An international exploration of governance, food systems, and environmental justice.* PhD thesis, Ghent University.

Putnam, R. (2001). *Bowling alone: The collapse and revival of American community.* Simon and Schuster Paperbacks.

Putnam, R. (Ed.). (2004). *Democracies in flux: The evolution of social capital in contemporary society.* Oxford University Press.

Šajn, N. (2020). *Sustainable consumption. Helping consumers make eco-friendly choices.* Briefing, European Parliamentary Research Service, European Union. Retrieved on March 23, from https://www.europarl.europa.eu/RegData/etudes/BRIE/2020/659295/EPRS_BRI(2020)6592 95_EN.pdf

Schwartz, S. H. (1992). Universals in the content and structure of values: Theoretical advances and empirical tests in 20 countries. *Advances in Experimental Social Psychology, 25,* 1–65.

SDSN & IEEP. (2019). *The 2019 Europe sustainable development report.* Sustainable Development Solutions Network and Institute for European Environmental Policy

SofProject. (2019). *Vision for Sofia. Report on people.* Retrieved on June 12, 2021, from https://vizia.sofia.bg/wp-content/uploads/2018/01/Доклад_Хора.pdf (In Bulgarian)

Srinivas, H. (2015). *Sustainable development: Concepts.* GDRC Research Output E-008. Global Development Research Center. Retrieved on June 21, 2021, from http://www.gdrc.org/sustdev/concepts.html

Tilkidzhiev, N. (2009). Trust and well-being. In N. Tilkidzhiev & L. Dimova (Eds.), *Well-being and trust. Bulgaria in Europe? Comparative analysis upon the European Social Survey 2006/2009* (pp. 33–62). Iztok-Zapad (In Bulgarian).

Tóth, A., Duži, B., Vávra, J., Supuka, J., Bihunová, M., Halajová, D., Martinát, S., & Nováková, E. (2018). Changing patterns of allotment gardening in the Czech Republic and Slovakia. *Nature and Culture, 13*(1), 161–188. https://doi.org/10.3167/nc.2018.130108

Trudel, R. (2019). Sustainable consumer behavior. *Consumer Psychology Review, 2*(1), 85–96. https://doi.org/10.1002/arcp.1045

UN. (2015). *Transforming our world: The 2030 agenda for sustainable development, A/RES/70/1.* Retrieved on June 21, 2021, from https://sdgs.un.org/2030agenda

UNEP. (2002). UNEP contribution to framework on promoting sustainable consumption and production patterns. In: *Draft working paper, United Nations environment program.* Division for Technology, Industry and Economics.

UNEP. (2015a). *Sustainable consumption and production. A handbook for policymakers.* Retrieved on June 21, 2021, from https://sustainabledevelopment.un.org/content/documents/1951 Sustainable Consumption.pdf

UNEP. (2015b). *Sustainable consumption and production and the SDGs.* Retrieved on 233 April, 2021, from Sustainable_consumption_and_production_and_the_SDGs_UNEP_Post_2015_Note_2-2014sustainable_consumption_and_production_and_the_SDG_english.pdf.pdf

UNEP. (2021). *Sustainable consumption and production policies. Online source.* Retrieved on 12 June, 2021, from https://www.unep.org/explore-topics/resource-efficiency/what-we-do/sustainable-consumption-and-production-policies

Veen, E., & Eiter, S. (2018). Vegetables and social relations in Norway and the Netherlands. A comparative analysis of urban allotment gardeners. *Nature and Culture, 13*(1), 135–160. https://doi.org/10.3167/nc.2018.130107

Vergragt, P. J., et al. (2016). Transitions to sustainable consumption and production in cities. *Journal of Cleaner Production, 134*, 1–12. https://doi.org/10.1016/j.jclepro.2016.05.050

White, K., Habib, R., & Hardisty, D. J. (2019). How to SHIFT consumer behaviors to be more sustainable: A literature review and guiding framework. *Journal of Marketing, 83*(3), 22–49. https://doi.org/10.1177/0022242919825649

World Business Council for Sustainable Development. (2011). *A vision for sustainable consumption. Innovation, collaboration, and the management of choice*. Sustainable Consumption and Value Chain. Retrieved on May 28, 2021, from https://docs.wbcsd.org/2011/10/AVisionForSustainableConsumption.pdf

Chapter 5
Economic Dimensions

Zornitsa Stoyanova and Galina Koleva

Abstract The chapter presents economic aspects of urban agriculture at different levels. The benefits appear on the farm and household levels, as well as on the city and macro levels. At the farm level, urban agriculture generates income and production of clean heathy food for market and self-consumption. At the city and macro level, urban agriculture creates employment and provides job opportunities. The potential of urban agriculture to promote the development of local food business is considered in the chapter due to people's orientation to healthy lifestyles and thinking about the food they consume and the environment they live in. The economic effects of urban agriculture for people and society are also an issue of the analysis. Some of the main lessons learned are connected with the following ideas: (i) urban farmers are a provider of quality niche market products; (ii) small urban farmers are important actors in short food chains; (iii) due to the direct contacts with their clients, urban farmers have more opportunities to trade through informal distribution channels, acting in the grey economy; (iv) initiatives from local authorities and targeted policy support are needed to create the conditions and perspective for sustainable urban agriculture.

5.1 Economic Impacts of Urban Agriculture at Different Levels

Literature pertaining to the concept of urban agriculture discusses its economic aspects at different levels. Research identifies economic impacts as those on farms and households. Other papers define the economic impacts of urban agriculture as effects at the city and macro levels. In order to understand the economic impacts of

Z. Stoyanova (✉)
University of National and World Economy, Sofia, Bulgaria
e-mail: zstoyanova@unwe.bg

G. Koleva
Institute of Philosophy and Sociology, Bulgarian Academy of Sciences, Sofia, Bulgaria

urban agriculture, a number of effects such as income generation, job creation, market orientation, saving on food costs, etc. must be analysed. The assessment of the economic impact at different levels varies significantly depending on: (1) the indicators used and (2) the objectives of the study.

5.1.1   Impacts at Farm and Household Level

The economic aspects of urban agriculture at farm and household level emerge most evidently when high-demand products are produced and when they have comparative advantages. At the household level, the effects are linked to the direct economic benefits and costs of urban households involved in agricultural production, including: self-employment, additional income from sales of surplus produce, savings on food and healthcare costs, agricultural exchange for others goods, etc.

Important factors affecting the impact of urban agriculture on the farm and its income are the degree of market orientation, the size of production, the availability of labour in the household, the choice of crops and animals, the availability and cost of basic resources, the possibility of irrigation, available technology and capital, access to markets and price security, ability to store, process and preserve products (Danso et al., 2003).

With regard to the economic aspects of urban agriculture at farm and household level, Kinkese and Pride (2017) link the effects to economic savings on food. Farm-produced food is consumed within the household, thus reducing the household spending on food, and they depend on the lower cost of self-production of food.

Another type of economic benefit of urban agriculture is that it serves as a source of income from the sale of crops. Simatele and Binns (2008) consider that the benefits of urban agriculture to income generation are most significant among poor people because most of them have smaller income and assets. Sources of income can also be renting out land and production sharing. However, urban agriculture can lead to a loss of household incomes and an increase in household food costs in the event of poor harvests due to adverse weather conditions such as floods, droughts, natural disasters, etc. (Simatele et al., 2012).

In this regard, Jamal and Mortez (2014) identified two groups of urban farmers in view of their motivation to practice urban agriculture. The former are people engaged in activities to meet their family's needs, and the latter includes those who view their activity as business with profit and intent (Jamal & Mortez, 2014). A similar categorization (urban gardening and urban farming) is also introduced by Lohrberg et al. (2016).

Research evaluating the economic aspects of urban agriculture also considers the economic benefits of this activity relevant to waste management. Smit and Nasr (1992) point out that the use of waste from urban agriculture could help overcome the challenges to waste management. The economic effects are, on the one hand, for households that compost and return bio-waste to the soil, and on the other, the attitudes of farmers and consumers to rationalizing consumption of farm food, regardless of its appearance, in order to optimally utilize food resources. Cofie et al. (2006) hold that urban agriculture can contribute by turning urban waste into productive resources. This could be compost production, vermiculture, irrigation with wastewater. On the other hand, urban farms produce bio-waste that, apart from being used in agricultural holdings, could also be sold or exchanged. Drechsel and Kunze (1999) consider that urban agriculture could bring down the cost of waste disposal by providing nutrient recycling of organic wastes.

5.1.2 Impacts on the City Level

Rapid urbanization requires additional jobs and suitable living conditions in the cities. On the city level, when assessing the economic impact of urban agriculture, the support to urban farmers for training, quality control and other activities on which farmers do not spend enough have to be considered. According to Fleury and Ba (2005), the positive effects of urban agriculture add value to the city, such as increased income or reduced food and waste disposal costs. They also define the positive and negative effects of urban agriculture, such as: waste recycling, less health problems, including those resulting from better nutrition of the poor in the cities, preserving the landscape, as well as water pollution caused by agrochemicals, erosion, etc. Urban farms can occupy unused space and abandoned and desolate land, which reduces municipalities' costs of maintaining the territory (Hallett et al., 2017).

The indirect costs and benefits of urban agriculture for the city, which can have an impact on the social, health and environmental condition of the urban population, must also be taken into account. According to Garnett (2000), local authorities recognize the potential of urban agriculture to include specific vulnerable groups in the socio-economic life of the city and their role in reducing poverty. As a consequence, support for urban farmers with access to municipal land, training, technical assistance and investment and marketing support for agricultural production and processing enterprises has been put in place in many regions.

According to FAO (2007), on the city level urban agriculture has impacts on the local economic development via income generation, developing micro-enterprises and employment. Urban agriculture also contributes significantly to ensuring food security and nutrition for the urban population, especially for the poor in the city. It also ensures social safety in times of economic or political crisis and can be used as a strategy to promote the social inclusion of disadvantaged people, leading to revitalization of the community. On a city level, urban agriculture contributes to the management of the urban environment, turning waste into resources, contributing to a better urban climate and managing the urban landscape.

5.1.3 Impacts on the Macro Level

The benefits of UA on the macro level are determined on the basis of the contribution of urban agriculture to gross domestic product (GDP) and its impact on the effectiveness of the national food system. Kinkese and Pride (2017) point out that urban agriculture creates jobs and is a source of employment. Landowners hire either seasonal or full-time employees depending on labour needs. Hunold et al. (2017) consider that the contribution of urban agriculture to the achievement of economic development goals such as raising capital, income generation, job creation is limited. The survey they conducted among farmers indicated that the respondents considered that urban agriculture is not economically viable with regard to the economic aspects of urban agriculture. Opinions on the potential for economic benefits of urban agriculture range from views that urban agriculture will not continue to be cost-effective in the future, to views that the economic benefits of urban agriculture may increase in a more favourable financial and political environment (Hunold et al., 2017).

Urban agriculture has the potential to stimulate the development of local economies in developing countries, providing both greater food security and significant job opportunities (Agbonlahor et al., 2007). Nugent (2003) states that the major macroeconomic effects of urban agriculture are related to the provision of food of value for relatively poor urban residents, lower food prices and increased food security. In this respect, urban agriculture has the potential to diversify the economic options and the access of citizens to food resources.

Urban agriculture also supports the development of the local economy by providing small investments to produce the necessary agricultural products, processing, marketing and services such as animal welfare, transport and self-employment across the community. In addition to the economic benefits for producers, urban agriculture can contribute to the creation of new businesses and job opportunities in and around cities. These opportunities, including fertilizer production, seed and tool production, marketing and distribution, processing, lead to the development of small businesses to create the necessary factors for agricultural production (Jamal & Mortez, 2014).

Keeping in mind the views on the degree of impact of urban agriculture at the three levels and shared sceptical views about the significance of this activity in economic terms, the potential contribution of urban agriculture to the development of farms, the city and the national economy in economic terms is summarized in Fig. 5.1.

5.2 The Potential of Urban Agriculture to Promote the Development of Local Food Business

More and more people are interested in healthy lifestyles and are becoming aware of the food they consume and the environment they live in. It may sound trivial, but food is the basis for human health – physical, mental, social.

What food we produce and consume is an issue with multiple projections, both depending on individuals' lifestyles and quality of life, and in the economic and socio-cultural context of social development. Food "is not just a collection of products based on statistical and dietary requirements, it is at the same time a communication system, a body of images, a passage for use and behaviour" (Barthes, 1961).

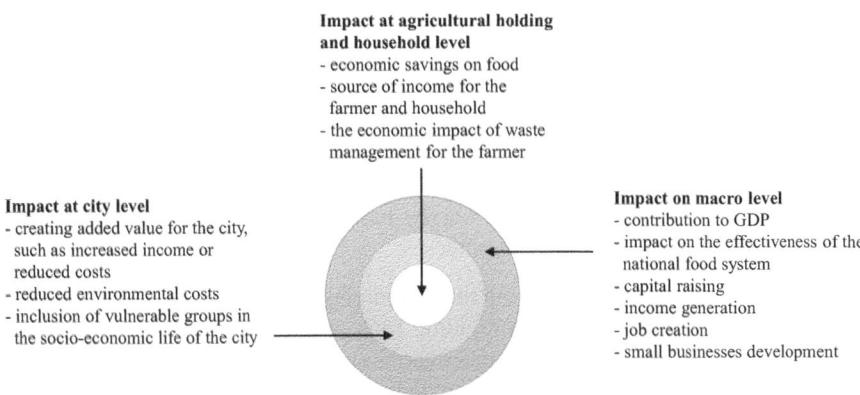

Impact at agricultural holding and household level
- economic savings on food
- source of income for the farmer and household
- the economic impact of waste management for the farmer

Impact at city level
- creating added value for the city, such as increased income or reduced costs
- reduced environmental costs
- inclusion of vulnerable groups in the socio-economic life of the city

Impact on macro level
- contribution to GDP
- impact on the effectiveness of the national food system
- capital raising
- income generation
- job creation
- small businesses development

Fig. 5.1 Potential economic impacts of urban agriculture at different levels

Food is part of the social life of the city, it has a cultural dimension, a saturation with cultural practices, and this is connected, on the one hand, with opportunities for a significant change in attitude towards it, as well as with impulses to rethink concepts such as authenticity, locality, identity (Petrova, 2014). On the other hand, the production of seasonal food by small urban producers, including in Sofia Municipality, is bound by concepts and labelling that attach higher value to its local origin – besides being fresh and freshly produced, it is *clean* (without the use of artificial fertilizers, chemicals), *tasty* (choice of varieties), *fair* (commitment and fair price for the farmers' labour).[1] The issue of local and global dimensions regarding food production as part of the food chain is a contested concept itself. For example, is it "local" food if seeds from the other end of the world are used to produce it (Gaspard, 2017)? In this chapter, we limit ourselves to the understanding of local food as food produced in a particular locality.

The views shared during the interviews with stakeholders in our survey have unveiled interesting storylines for interpretation regarding the food produced – it is evaluated not only as a product and a means to meet basic human needs, but also as a multifaceted socio-cultural attitude with different projections – food as a socializing factor and a means of change, food as an experience, food as heritage, food as connection, solidarity and ethics, food as waste.

The results from the project support the thesis that the potential of urban agriculture goes far beyond its narrow understanding of "food production", but rather it is positioned as a necessary and important element of the urban environment, contributes to solving economic, social and environmental problems, including in relation to important issues of society such as climate change, inequalities and poverty, social responsibility and community support, quality of life and human happiness.

---

[1] Shared in one of the interviews (a representative of Slow Food for Bulgaria) in the framework of the project "Urban Agriculture as a Strategy for Improving the Quality of Life of Urban Communities", funded by the NSF.

Theses and conclusions are based on the analysis of information obtained during the study mainly[2] through interviews with market-oriented urban farmers who grow seasonal vegetables, spices, fruits and dairy products, mainly in the suburbs of Sofia Municipality. Market-oriented producers are one group of stakeholders that have been in the focus of qualitative research with their role as business-oriented actors in urban agriculture.[3]

In particular, the results from the survey based on qualitative and quantitative methods supported the following hypotheses:

- Urban agriculture practised within the boundaries of Sofia Municipality does not have so much importance for providing food and economic benefits (with the exception of existing business-oriented farms and initiatives), but it rather focuses on educational, social communication and value issues, and offers solutions for environmentally friendly, nature friendly and sustainable development.[4]
- Small producers and family farms in the suburbs of Sofia offer a wide range of products on the market (through farmers' markets, e-commerce, informal consumer networks and face-to-face sales) – from fresh organic vegetables, spices, fruits, honey, nuts and animal products, to a wide range of plant-based and dairy processed foods. The producer Chilli Hills is an example of a fast-developing, end-to-end business model – from growing various varieties of hot peppers, vegetables and spices, through their processing into sauces and other canned products in their own production plant, to marketing, distribution and sales on the national and international markets, including in its own brand shop in the central part of Sofia, where both fresh and processed proprietary products are available (Georgieva, 2018).
- Urban agriculture as a market realization is in a process of gradually entering and strengthening the emerging *market niche of environmentally-friendly, fresh and local food,* for which there is steady demand and growing expectations to "work" towards healthy food and a healthy lifestyle. Consumption of quality food (environmentally friendly, established or sourced from local farms) is a basic

---

[2]Relevant information from interviews, focus group discussions and expert assessments from other project stakeholders is also analysed, as well as from additional studies.

[3]Five interviews were conducted with farmers focused on the marketing of seasonal vegetables, spices, fruits and dairy products, produced mainly in the suburban areas of Sofia Municipality. Three of them (Versa Natura, Chilli Hills, baby vegetable grower) run successful organic vegetable businesses, create their own products and model, rely on their own network of people and business associates; one respondent (a goat and sheep breeder in Lozen) produces dairy products (yoghurt, cheese and meat), selling them primarily to regular customers seeking pure, natural foods with which he has informal trust and respect; director of a field station (Training and Experimental Field Station of the University of Forestry, Vrazhdebna), which is licensed as an agricultural producing legal body where, in addition to carrying out agricultural activities for educational and research purposes, direct sales of milk to consumers from Sofia are offered (less frequently live animals) as well as selling milk to a processing plant on a contractual basis.

[4]More detailed information on the results of the qualitative research can be found on the project website (ISSK, 2020).

value for all surveyed persons – 96% of them recognize and evaluate it as "very important" or "important". The benefits of urban agriculture, considered in the dimensions of "production of organic products" and "opportunity for healthier eating" are evaluated as important respectively by 63.9% (477 surveyed people) and 56.3 (420 surveyed people) of the respondents.

- We also found support for the hypothesis that small local food producers have an important role in the establishment of an integrated economic system that not only contributes to employment, better income, entrepreneurship and business, but it builds a broader socio-cultural background for economic development by supporting traditions and local identity, upholding prestige and uniqueness, mutual respect and building trust, recognition and embeddedness in the environment and community. Some urban farmers operating in the local market function as small production enterprises and holdings with the intrinsic characteristics and advantages of the "small" ones that make them important and necessary for the local market and society relying on family labour, better flexibility and adaptability to the environment and change; good local knowledge, supporters of local culture and traditions, including varieties and biodiversity, contributing to the diversity of products and the diversity of local cuisine.
- The role of urban agriculture and urban producers is significant in support of the promotion of a consumer's model alternative to mass and conventional consumption; a model contributing to the promotion of another type of values on the one hand, and a responsible and supportive attitude towards the local environment, on the other. Small-scale urban farmers are an important entity in short food supply chains, but they are at the same time a factor in their development as sustainable supply and consumer channels. These are the fast-growing farmers' markets, traditional and natural food festivals, on-line delivery networks for organic products, specialized health food stores. These relatively new consumer-friendly models reinforce environmental attitudes and values, responsible attitudes and behaviours to nature and the urban environment, value for nature and food-friendly origin, and more responsible treatment of waste, composting and recycling.
- The role of farmers' markets as part of the local food chain and an important partner of urban farmers and other businesses in the niche of local food is highlighted as necessary for the development of the local food business. There are currently six farmers' markets in Sofia, initiated by Hrancoop – Sofia, a cooperative of producers and consumers. It organizes weekly farmers' markets in Sofia as well as weekly deliveries via the internet to fulfil the mission of supporting small-scale organic food producers – certified and non-certified organic, but with the food production cycle controlled by the cooperative. In addition to promoting the small local producer and enabling direct contact with consumers, the farmers' market stands behind the producer as a guarantor of the quality and origin of the products offered. The selection of producers to join Hrancoop, respectively, to market their produce, is based on clear principles and regulation, a *"5-step model of selection and interprofessional control"*, as well as *"farm mapping – what and how they produce, what natural resources they have*

at their disposal (a river, a forest nearby, etc.)" and further generates consumer confidence that the food offered is not from random, illegitimate producers. As our respondents from farmers' markets (Versa Natura, Chilli Hills) point out, the farmers' market for them is a secure, regular market where they gain experience in communicating directly with customers and peers, *"they explain how they grow their produce, what they do, ways to store produce, recipes to use"* and *"very quickly the producer gets information about what and how he should do, and at what price to offer a product. There is feedback – after launching a product, a producer can figure out if it will be successful as early as the first week"* (coordinator of a farmer's market). An important conclusion is that the farmers' market "brings to light" the activity and the corresponding legitimacy by charging fees and demanding social responsibility, which is not valid for all small local producers remaining in the shadow economy. In addition, apart from organization, logistics and marketing services for farmers, Hrancoop and their organized farmers' markets cooperate with other small businesses such as bakeries, organic-shops, stands for dairy and meat products to rent stands and other services and facilities.

- Production and demand of local food is also linked to the strengthening non-commercial attitudes and values aimed at social cohesion and inclusion, such as developing solidarity and mutual assistance, sharing and community empowerment, building sensitivity to inequalities and vulnerable groups.

A number of projects and studies have focused on the role and place of urban farmers in view of food security and, with scientific arguments, testify to the importance of local farms for food security and its support of local food movements (Gaspard, 2018). One of the study objects of the trans-European project "Urban Agriculture Europe[5]" is the dimension of urban agriculture. The participants in the Working Group on Entrepreneurial Models of Urban Agriculture within this project analysed and compared urban farms and projects in the context of their innovativeness and adaptability to the urban environment, their involvement in the economic system and their effects on economy and society. One conclusion they reached was that "urban farms have the potential to be the 'hidden champions' of a green urban development strategy" (Van der Schans et al., 2016: 82). The applied CANVAS business model (Henriksen et al., 2012, Van der Schans, 2010), adapted to describe and analyse farms in various European cities, provides an overview and comparative analysis of key success factors, obstacles and barriers, the potential to generate business ideas and innovation for urban agriculture in Europe (Van der Schans et al., 2016: 82–102, Pölling et al., 2017). As other studies have shown, urban and suburban farms adapt and benefit from urban environments by successfully implementing high value-added activities and product-oriented strategies, niche products based on organic production, as well as through diversification of activities, including a wide range of non-agricultural primary agricultural activities and leisure, hobby, health, education, cultural and nature-related activities (Brayant et al., 2013, Zasada, 2011).

---

[5]COST action Urban Agriculture in Europe (2012–2016) is a networking project funded by the European Cooperation for Science and Technology (COST).

The research goals and objectives set out in the project "Urban Agriculture as a Strategy for Improving the Quality of Life of Urban Communities" do not place a particular focus on the detailed characterization of suburban farms, their economic and business characteristics, insofar as the project focus is placed on urban agricultural potential to generate social, economic and environmental benefits and enhance the quality of life of citizens through urban farming practices, including through the production of fresh local food. This limits the collected data to self-assessment and self-identification in terms of role, relevance, public benefit, as well as the public perception of urban farmers and their production. The information received from our key respondents reveals some of the opportunities that local urban farmers create through their activities and products to meet needs, to stimulate other industries, to develop new business niches, and to engage and interact with other businesses.

The urban market-oriented farmers interviewed perceive their activity as a business – they produce, have production costs, offer their produce on the market and have income, regular customers, some of them work with suppliers and distributors. Farming is the main employment that ensures their livelihood; a job they love to do and they are looking for profit maximization.

> Yes, it's a business because we want to maximize our revenue. We are not able to cover all the expenses, but we have always wanted to have our own income and profit (. . .) we live on our activity (Sofia urban farmer).

The results show that the farms surveyed are successfully integrated and well recognized on the market and each producer finds their own way to connect with consumers. This happens through various channels:

- through farmers' markets organized by Hrancoop with the assistance of regional administrations;
- through "direct access to 3-5 clients";
- through their own marketing strategy and established distribution network;
- through direct sales to regular customers who contact the producer through social networks or through acquaintances and friends;
- through contractual relations with other market entities, as well as through direct sales.

An example of an entity that uses long-term contracts and direct sales is the Experimental and Training Field of the Forestry University, presented in Box 5.1.

Box 5.1: "Vrazhdebna" Training and Experimental Field Station
"Vrazhdebna" Training and Experimental Field Station: a model of urban agriculture in research and education with economic activity involved

The field station was established in the period 1961–1963 for the needs of the Agricultural Academy in Bulgaria and its main mission is the promotion of research and education. Over the years, the intertwining educational, scientific

(continued)

Box 5.1 (continued)

and agricultural activities have turned the field station into an economic entity involved in the city's economic life – both in the market of food produced in the city and in the development of the market of urban agriculture products (production and trade in seeds, tools, etc.).

The field station is a registered agricultural producer with all responsibilities and rights arising thereof, including the right to sell the final product. The activities of the "Vrazhdebna" field station are financed mainly by the budget of the University of Forestry, but additional proceeds are generated from its output – milk and plant products. Part of the feed given to animals is produced by a feed mill in exchange for raw material – cereals.

The field station farms cows and calves, sheep and goats, pigs and rabbits, chickens, bees. Crop activity includes a wide variety of crops – cereals and oilseeds, over 500 fruit trees, all kinds of vegetables. The area of the field station includes pastures, perennial lawns for feed (alfalfa), natural meadows, an arboretum, ornamental nursery, greenhouses, vineyards with a collection of 54 varieties.

The activity of the field station has a very high contribution to the development of sustainable agriculture. Water is used as sustainably as possible – with a drip irrigation system using groundwater. Locally-produced fertilizers account for 100% of the total quantities of soil improvers used. Much of the biomass produced on site is composted.

Research experiments are carried out in the field station, for example. An experiment for roof urban agriculture is currently underway, testing the effect of the urban environment on the pollution of edible plants.

The motivation of the urban farmers surveyed is based on attitudes and beliefs of the acquired ecological values and attitude towards the environment, a personal connection and relationship with the land and agricultural work, shaped in a family and social environment, the perception that clean food and healthy nutrition are the basis for a higher quality of life. On the other hand, motivational choices are realized, measured and activated through "entrepreneurial endowments", i.e. through personal qualities such as a desire for self-development and self-fulfillment, a tendency to take risks, and by having an entrepreneurial flair for promising segments in consumer interests and expectations. An important motivational impulse is the understanding that successful marketing depends on direct contact with consumers who want and expect to be well informed about the origin of food, the conditions and methods of production, how it is promoted and delivered. An important and stimulating intermediary role is played by social networks to enhance the interest and confidence in small local producers as more reliable suppliers of organic produce, compared to large retail chains that offer, as a rule, agricultural output produced under conventional methods and targeted at the mass market. It could be summarized that the motivational choice for job creation related to the prospective market niche in the search for clean local food is a sustainable factor for the development of a local food business.

The subjective evaluation and (self-)identification of the products is another important factor contributing to a sustainable positioning in the local food system and market. The producers surveyed identify the qualities and uniqueness of their production, through the methods of growing it – organic food produced without chemicals and preparations; produce grown on clean land and by controlling all the parameters at the stages of production. Farmers share a view that they use only organic fertilizers and organic products, compost, and use products only from specialized stores. The products thus produced are in higher demand among customers, oriented towards better quality food, responsible for their own health and striving for a better quality of life.

The increasing demand for seasonal production from local producers contributes to the sustainability of their businesses. All of the farmers surveyed have increased their production to some degree over the past year. The small family vegetable farm Versa Natura has doubled its output (compared to the previous year) and has built a second greenhouse for early vegetables; Chilli Hills, again a family business producing chilli peppers and hot sauces closed the "seed-to-shelf cycle" with seasonal and in-house production; a baby vegetable producer is steadily working for several restaurants with day-to-day supplies. At present, after the closing of restaurants due to COVID-19 measures, he has re-directed his produce to the farmers' markets and a large wholesale trader. Direct contacts in the sale of products contribute to an informed choice and traceability of the product, as well as its identification with the producer, which is especially appreciated by customers and consumers.

> People know our production not as local, but as first-hand production, from a producer, so they look for us, they are not forced to go through ten distributors to get the product they want (Chilli Hills).

Two of the farmers say that the success of produce and a brand is due to the *"the brand's uniqueness"* and that they offer something that no one else offers. Such self-assessment for product identification is also in support of the thesis that the small producer's advantage meets emerging needs and market demand for something different, non-mass produced, more exotic or very traditional but absent from the market, as well as gourmet style, and thus contributes to the differentiation of the local food system. The spectrum of users of food produced by urban farmers includes the producers and their families, direct consumers who are *"usually young people, educated, financially well-off, with healthy life"*, as well other food business operators – distributors, shops, restaurants, slaughterhouses, dairies and food businesses.

The analysis of the information from the study supports the important conclusion that urban farmers are successfully integrated into the niche market with a focus on healthy food and healthy lifestyles. This is a relatively new market niche that is evolving in response to the needs, expectations and growing demand for seasonal, locally grown, locally sourced food labelled as "natural, eco-friendly, organic, authentic". Producers use these "health claims" to promote the image of their products as "more genuine, useful and delicious", and some also rate it as a *"unique or specific product"* for more selected tastes.

Urban agricultural producers evaluate direct sales as a great advantage for them. They could meet their customers in person, advertise their products, gain trust with their competence and openness to communicate, and accordingly have the economic benefits of the image of the products, which is already not anonymous for the user. This helps them strengthen their position in the market segment.

In conclusion, discussing the economic benefits of urban farmers' activity and based on the analysis of qualitative survey information about the profile of market-oriented producers, we can position these benefits in a broader context as follows:

- Farmers who work in the field of urban agriculture through their activities develop the potential to promote the expansion and market positioning of the local food business, which is promising and influences both quality of life and sustainable development, in economic and in social aspects. More and more people are interested in a shared understanding of a healthy and environmentally friendly life, as well as in support of local producers.
- Local food offers opportunities for synergy between new and established businesses by supplementing, upgrading production and services, by tightening and closing production chains that target specific high-demand customers.
- Urban agriculture contributes to the diversification of business services – an example of this is the trade with healthy food in restaurants that do not just offer food, but put a focus on its qualities (gourmet, "real products"), vegan restaurants, recreation services, culinary events with a focus on traditions and customs, farmers' markets, educational initiatives based on food and culinary skills.

- Urban farmers are embedded in the urban economic and social environment which vests them with different capacities and responsibilities that go beyond production. On the other hand, the impact of the city on farms and agriculture is complex and is associated with new opportunities, niches for the development and diversification of agricultural activities.

5.3 Economic Effects of Urban Agriculture for People and Society

Respondents in our study define economic benefits of UA as benefits for the individual and benefits for society.[6] The economic impact of urban agriculture is not defined as significant either in the long term or in the short term.

5.3.1 Economic Effects at Farm Level

Most of the respondents involved in the qualitative study (25 in-depth interviews, five focus group discussions and five expert interviews) consider economic benefits to be related to possibilities urban agriculture creates for subsistence and cost savings from the products consumed by the household. Individual economic benefits are linked to food production and the saving of financial resources. They state that for the agricultural producer and the household, the economic impact may be quite substantial but it may be small on the city level. Others associate the economic benefits with the household's saving on food cost as a result of the production.

> You will not pay money for things that are in the store. If people have a garden of their own, they will not pay for certain vegetables that they can grow themselves and they can also have the revenue that is generated when they sell their produce (focus group participant).

Some respondents involved in the qualitative study believe that their economic performance is very good, especially during the season of production and the economic dimension is expressed in financial income and employment.

> The results are very good. From a financial point of view, when we are in season and we have produced, we, the family, live on what we offer in the markets, we cultivate three acres. It really is justified, it makes sense from a financial point of view – economic benefits, employment (family commercial urban farmer).

---

[6]The qualitative study in the frame of the project serves as an empirical basis for the analysis and conclusions. It includes 25 in-depth interviews; five focus group discussions and five expert interviews.

Some respondents who were part of the qualitative study and are not commercial farmers themselves consider the economic benefits of urban agriculture to be negligible because of the small volume. These respondents do not associate the purpose of urban agriculture with economic profit, because the quantity of production is small and is too limited to provide the livelihood of urban residents. According to them, no significant economic dimension can arise from urban agriculture, since products are used mainly for their own consumption, with the exception of the surveyed market-oriented farmers who produce for the market. The reported satisfaction of the practising commercial farmers with the revenue they get from their practice indicates that this opinion does not reflect the economic reality of market-oriented UA in Sofia.

Similarly, a common perception of non-commercial urban farmers is that urban agriculture has no economic dimension, and its goal is rather educational and social – to get people to know how food is produced, where food comes from, to become more creative, to create, socialize and have fun.

Despite the widespread view that the economic benefits of urban agriculture are negligible, some participants in the qualitative study consider that the economic benefits may increase, given the increasing demand for high-quality and locally produced food. They share the opinion that, although there are no economic effects at this stage, they can arise if urban farming practices become widespread and the people prefer *"food they buy to be produced in the city or around the city"* (coordinator of a farmer's market).

Another indirect economic aspect of urban agriculture for the individual agricultural producer is related to the production of clean food and the improvement of health. The economic benefits of urban agriculture are due to the consumption of healthy food, which has been shown to have positive effects on health and thus leads to lower costs for healthcare, medicines and therapy.

The quantitative survey[7] conducted shows that, at the farm and household level, income is not a leading motive for urban farming. Still, there is a slight correlation between respondents' incomes and their past or present experience with urban agriculture (Cramer $V^2 = 0.143$; Chi-Square$X^2 = 0.008$), presented in Fig. 5.2.

The motives for practising urban and peri-urban agriculture are of significant importance and they enable us to draw conclusions about the economic benefits of this activity. The economic motivation to practise urban and peri-urban agriculture is significant for a small number of respondents. Only 2.5% of respondents who practise in the city state that the motive for practising this activity is to earn extra income and 17.4% of urban agricultural producers state that a motive for them is to meet basic nutritional needs (See Fig. 4.1). The low share of respondents that have given these answers indicates that the basic motive to practise urban and peri-urban agriculture is not of an economic nature.

---

[7]The quantitative research design presented in Chap. 3 serves as an empirical basis for this analysis.

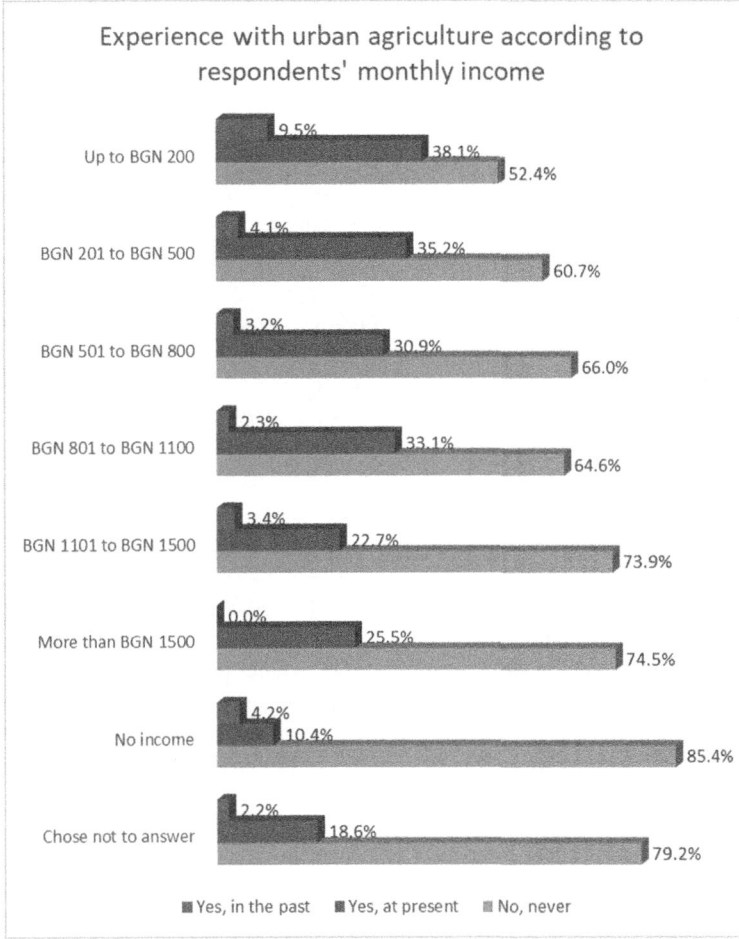

Fig. 5.2 Experience with urban agriculture according to the income of respondents

5.3.2 Economic Effects at City and Macro Level

Respondents agree that the economic impact of urban agriculture on the city and society as a whole is not large, since the share of the total production of urban agriculture in the city is relatively insignificant and, in this respect, urban agriculture cannot solve the economic problems of the region such as food security, employment, poverty. At the same time, some of the respondents indicated potential economic benefits that are significant for society. They are associated with composting of waste. The overall economic effect is *"saving from the disposal of plant residues, garden and park waste, which save the municipality additional funds for transportation to the landfill of the municipal waste collection system"*.

According to the respondent, a household waste tax may be linked to whether the household composts and accordingly means the municipality has to transport less waste. Furthermore, the economic impact on society in view of fuel economy is related to the elimination of the need for long-distance transport and the preservation of road infrastructure.

Economic effects on people and society include the benefits shared by the respondents in terms of job creation and job opportunities, turnover and income. Respondents consider that the main economic aspect related to improving the quality of life is the creation of employment. Opinions differ about the greatest benefit of urban agriculture on various groups. The analysis of the respondents' view involved in the qualitative survey shows that these practices would encourage retirees and the unemployed people to be motivated to produce their own food and become engaged in more intensive social communication. Others cite minority groups as a targeted stakeholder as UA can lead to employment of minority groups, which could have a very strong economic and social impact if *"these people are engaged in labour"* (representative of a social kitchen and civil organization supporting disadvantaged citizens).

Some respondents consider the economic benefits of urban agriculture by linking them to the production of better quality products at a lower cost. According to them, from a societal perspective, the economic benefit is that better quality products are created because the product life cycle traceability is facilitated and control of the production process can be carried out. There is a direct producer-consumer connection, which leads to a lack of surcharges and a lower cost of production. The price is relatively low because it is purchased directly from the producer and the supply and communication chain is short. Smaller producers produce higher quality products as they produce in smaller volumes.

5.3.3   Factors That Influence Economic Aspects of Urban Agriculture

Regardless of the level of economic effects of urban agriculture, a number of factors such as the degree of market orientation, the size of the holding, the availability of workforce in the household, the choice of crops and animals, availability and cost of basic resources (and in particular the use of local resources such as organic waste and wastewater); the possibility of irrigation, available technology and capital, access to markets and price assurance, the ability to store, process and preserve products, influence the strength of the economic benefits of urban agriculture. Some of the significant factors influencing the identification of economic aspects of urban agriculture are as follows:

- Respondents' access to natural resources

The respondents' replies on how they get access to natural resources, in particular land and water as a means of producing and carrying out their activity, show that they have different means of accessing these resources. Some of them use their own the land and water wells for production and irrigation. Others use leased municipal land and water resources close to the place of production. Some of the agricultural producers carry out urban agriculture on terrains of unknown status. Some of the respondents shared the view that they were not informed whether the municipality provided resources for urban agriculture. However, they note that under some projects (the TURAS project,[8] for example), Sofia Municipality has expressed its readiness to take over the maintenance of the irrigation facilities and to provide a market for the sale of the UA produce.

The opinions on the provision of land and water resources show that in most cases the water used for production activities is obtained from wells or collected rainwater is used, and the status of the land in terms of ownership is diverse. Respondents' opinions regarding the status of land depend on the forms of practising urban agriculture – market-oriented holding, communal garden, demonstration organic garden, educational garden. Most often, in business-oriented producers, it has a regulated status (own land or rented), and in the case of urban agriculture as a communal garden, in some cases, the status is unclear. In the cases where urban agriculture has an educational function, it is most often state land or is owned by the municipality.

- Financing

Financial support and incentives would motivate many people to focus on implementing urban farming practices. The interviewees consider that they need financial support to implement more projects for this type of activity. Regarding issues related to specific assistance and funding, respondents believe that there is a need for funding for ideas related to urban agriculture and that funding is needed for this type of activity, but support should be targeted after analysing the regulatory framework. They claim that *"artificial promotion will not help, but it will be useful to have an analysis of the regulatory environment in which specific initiatives have to be implemented."*

Some of the respondents consider that their initiatives have been implemented through voluntary participation and funding through donations, but this is not enough to ensure the sustainability of urban agriculture.

Some interviewees believe that financial support can be provided through tax breaks, preferential loans, financial support programmes at national and EU level.

- Distribution channels

---

[8] "TURAS: Transitioning towards Urban Resilience and Sustainability" was an EU funded project within the 7 FP for research, under grant agreement No 282834.

The surveyed commercial producers sell urban agriculture production through various distribution channels.

Some of them offer urban agriculture products to farmers' markets. Most often, they are organized by a particular company with the permission of the municipality *"to provide place for organization of a farmers' market"*, either on a formal or informal basis. Some of these markets operate on an ongoing basis, others are monthly or they are dedicated to a particular holiday (Easter, Christmas, etc.) It is civil organizations that initiate farmers' markets to support small producers who do not have shops and use the opportunity to sell their production otherwise. Respondents consider that the organization of the farmers' market has a number of benefits related to the satisfaction of people to visit this market, as well as satisfaction from contacts and communication between farmers on the market. The respondents mention the educational function of the farmers' market, where knowledge and advice about urban agriculture could be spread among farmers.

Other respondents prefer not to participate in farmers' markets. This is attributed to the physical inability of producers to participate on certain days of the week, others find it pointless as they have enough customers that they regularly supply.

Respondents receive information about farmers' markets from various sources. Some of them are indirectly informed by the municipality, others are informed by social networks. Producers are also informed of topics of interest to the market.

Some respondents use as a distribution channel "organic-shops and shops promoting local food", and these shops have a policy and concept to promote Bulgarian food and Bulgarian producers.

The third opportunity for urban agriculture producers to sell their products is direct delivery. They have established relationships with stores that directly deliver packaged production, alternatively, they deliver themselves or have the customers pick up their purchase on the spot from the farm. In these cases advertising is limited because quantities are small and there is no surplus left over that needs to be sold off.

- Marketing of production

Respondents also point at some difficulties with respect to marketing of urban agriculture production. On the one hand, the regulatory framework of Sofia Municipality does not refer to the concepts of urban agriculture and farmers' markets. In this regard, some of the respondents stated that the Ministry of Agriculture is making efforts to amend the regulations for markets with the aim to include the concept of "farmers' market" to explain better the establishment and organization of farmers' markets in the legislation. This will support agricultural producers in obtaining easier access to markets. In this connection, the following problems with respect to the sale of urban agriculture production are considered:

- Difficult access for small producers to the market due to heavy regulations and requirements. Respondents consider they need to *"be supported in bureaucratic burdens"*. It is necessary *"to have clear conditions as to how, under what circumstances, when and where such markets are organized"*. They say that in the regulations and documents of Sofia Municipality, information about the farmers' markets is limited. They share a view that farmers need more information about *"what is a farmers' market, who can organize it, the rules, the bodies to assist the process, etc."* There is no special policy of Sofia Municipality for the development of urban agriculture. The support, whenever such has been declared, is only in words and without large-scale action.

- Lack of infrastructure and technical parameters related to street market organization.
- The inability of relatively small producers to provide large volumes of production. The respondents consider that *"it would be difficult for farmers to produce every week and they do not have that attitude."*
- A disabling factor for commercial UA development is also the existing suspicion of some actors that there is an optimistic future for thriving urban farming in Sofia. There is still a general belief that this type of farming is yet to prove its viability. As an example, one of the producers said *"unfortunately, as every new thing, urban agriculture is connected with many problems and hurdles. The development of these ideas is very hard, I hope things change, but for the moment it's very difficult."*
- There is also a problem of lack of consumer trust in the farmer's markets and food selling points in general, because traditionally food markets in Sofia have been taken over by traders who are not producers themselves. While farmers' markets aim to offer a marketplace for genuine producers, and their existence is a priority in the European Common Agricultural Policy, Sofia Municipality has not so far reflected this priority in its own local policies. This has led to the situation where the vast majority of consumers do not understand the difference between farmers' markets and typical traders' markets which offer lower quality and lower priced products and thus they do not find the higher prices justified. On the other hand, the producers selling at farmers' markets are not united and they do not act in a strong collective manner, which further hinders the establishment of the farmers' markets as a sustainable channel for local food sales.
- Leading on from the previous point, there is a pressing need for the political will to recognize farmers' markets as a key channel for distributing high quality local food in the city. The role of the local authorities is very significant in overcoming the barriers to flourishing farmers' markets by introducing clear regulations for them. This would provide security for the farmers and the organizers of the markets who, at present, are at constant risk of cancelled market events and lost market opportunities at a whim of a local mayor or other official.

In conclusion, factors that influence the increasing economic benefits of urban agriculture are: (1) The availability of natural resources for urban agricultural producers as a source of production. There are diverse ways to make these resources available in the opinion of agricultural producers. (2) Financial support and incentives that would motivate many people to focus on urban farming practices. Financial support is needed to carry out more projects for this type of activity. The financial support has to be targeted after regulatory analysis. (3) Marketing of production through various distribution channels – farmers' markets, organic shops, stores that support and sell local food, direct delivery.

5.4   Lessons Learned

Economic aspects of urban agriculture could emerge on three levels: the agricultural holding and household, city and macro level.

At the farm level, urban agriculture generates income from the sale of production, production of food for self-consumption and job creation. The economic benefits are more for the economic entities which embark on urban agriculture. They are expressed in providing a livelihood and cost savings of products consumed by the household. The majority of respondents consider the economic benefits to be of least importance, because the amount of production is small, too limited to provide a livelihood to citizens. Nevertheless, they consider that there is still an impact on urban farming practices, which is economic in nature, especially for market-oriented urban farmers.

The economic aspect of urban agriculture at farm level is related to the production of clean food and the improvement of the health status of people that results from the consumption of healthy food, and this results in the indirect economic benefits because they save money on healthcare due to the healthier lifestyle.

At city and macro level, urban agriculture creates employment and provides job opportunities. It has the potential to create employment if agriculture is organized and structured as a business model that creates jobs. Urban agriculture can potentially provide employment opportunities for minority and disadvantaged groups, and this type of initiative and engaging these people would have a profound effect on them and the community, supporting efforts to overcome some problems related to poverty and social exclusion. The economic benefits of urban agriculture at city and macro level are not highly appreciated by the respondents, although there is a consensus that there are niches where providing specific business models for small-sized industries ensures employment and income for citizens. These models include mainly flexible part-time work to provide additional income, but also full-time farming, especially in peri-urban areas. The economic benefits of urban agriculture are also linked to the production of better quality products at a lower cost. From a societal perspective, the economic benefit is that better quality products are created because the product life cycle traceability is facilitated and control of the production process can be carried out. There is a direct connection between producer and consumer, and the supply and communication chain is short.

Some of the main conclusions and lessons learned are as follows:

- Urban farmers are a provider of quality niche market products emerging in Bulgaria around the demand and supply of healthy food and increasing awareness of a healthy lifestyle. Farmers from the suburbs of Sofia offer a wide range of products for market realization (through farmers' markets, online commerce, informal consumer networks and face-to-face sales) – from fresh organic vegetables, spices, fruits, honey, nuts and animal products, to a wide range of processed foods based on crops and dairy products. Local food is sought by both direct consumers and heterogeneous food business operators – distributors, shops, restaurants, processing plants. The local food business can influence the quality of life and sustainable economic and social development.

- Small urban farmers are an important player in short food supply chains. By taking advantage of them, farmers are at the same time a factor in establishing these types of chains as sustainable supply and consumer channels and as an important part of the local food chain. Producers who participate in the farmers' markets are convinced of their economic need and benefits, not only because they meet their customers there, but also because they bring agricultural business out of the grey economy, enhance its prestige, and work towards higher trust and respect for the local and high quality products sold at these markets.
- There are also unregistered urban farmers offering produce through informal distribution channels. There is a lack of control over their production, and generally they do not declare their turnover and income. These market-oriented farmers generally rely on "word of mouth"; "friends of my friends" and sell their products exclusively through informal relationships; most have regular customers, and some rely on occasional sales only. The issue of selling home-made foods free of quality control and beyond all registration and accountability requires special focus and research on the various categories of customers willing to obtain food outside the usual legal distribution channels. Our research does not focus on this information.
- Real support and initiatives from local authorities are needed to create conditions and prospects for sustainable urban agriculture, including conditions that stimulate the activities of local plant and animal food producers. Of particular importance to them are the support and initiatives of local authorities in the following areas: support for farmers, including creating, regulating and maintaining markets and facilitating access to them; construction and maintenance of infrastructure (roads, lighting); composting services; regular collection of waste; promoting the activities and production of small local producers and supporting the "prestige of urban agriculture".
- There is also no clear and structured food policy at the municipal level, covering hobby UA activities, socially or educationally-aimed UA initiatives, and commercial farmers. This makes the potential economic benefits of UA in Sofia difficult to make full use of.

References

Agbonlahor, M. U., Momoh, S., & Dipeolu, A. O. (2007). Urban vegetable crop production and production efficiency. *International Journal of Vegetable Science, 13*(2), 63–72. https://doi.org/10.1300/J512v13n02_06

Barthes, R. (1961). Pour une psycho-sociologie de l'alimentation contemporaine. *Annales. Histoire, Sciences Sociales, 16*, 977–986.

Brayant, C., Sanchez, N. C., Delusca, K., Daouda, O., & Sarr, A. (2013). Metropolitan vulnerability and strategic roles for peri-urban agricultural territories in the context of climate change and vulnerability. *Cuadernos de Geografía: Revista Colombiana de Geografía, 22*(2), 55–68. https://doi.org/10.15446/rcdg.v22n2.37016

Cofie, O., Adam-Bradford, A., & Drechsel, P. (2006). Recycling of urban organic waste for urban agriculture. In R. van Veenhuizen (Ed.), *Cities farming for the future. Urban agriculture for green and productive cities* (pp. 172–208). RUAF Foundation.

Danso, G. K., Drechsel, P., Akinbolu, S. S., & Gyiele, L. A. (2003). *Review of studies and literature on the profitability and sustainability of urban and peri-urban agriculture* (FAO Final Report (PR 25314)). IWMI.

Drechsel, P., & Kunze, D. (1999). *Synopsis from International Workshop on Urban and Peri-Urban Agriculture: Closing the Nutrient Cycle for Urban Food Security and Environmental Protection.* Retrieved June 12, 2021, from African Peri-Urban Workshop (cityfarmer.org)

FAO. (2007). Profitability and sustainability of urban and peri-urban agriculture, Agricultural management, marketing and finance. *Agricultural management, marketing and finance occasional paper, 19.* Retrieved June 12, 2021, from http://www.fao.org/3/a-a1471e.pdf

Fleury, A., & Ba, A. (2005). Multifunctionality and sustainability of urban agriculture. *Urban Agriculture Magazine, 15.* RUAF foundation. Retrieved June 12, 2021, from https://ruaf.org/document/urban-agriculture-magazine-no-15-multiple-functions-of-urban-agriculture/

Garnett, T. (2000). Urban agriculture in London: Rethinking our food economy. In M. Bakker, S. Gündel, U. Sabel-Koschella, & H. de Zeeuw (Eds.), *Growing cities, growing food: Urban agriculture on the policy agenda* (pp. 477–501). German Foundation for International Development.

Gaspard, A. (2017, September 28). *Going beyond local food.* Urban food futures. Retrieved June 12, 2021, from https://urbanfoodfutures.com/2017/09/28/going-beyond-local-food

Gaspard, A. (2018, April 26). Is local food good for the economy? Looking at the full picture. Urban food futures. Retrieved June 12, 2021, from https://urbanfoodfutures.com/2018/04/26/economic-impacts/

Georgieva, M. (2018, November 28). Chilli Hills Foods is expanding its production in a new workshop with a BGN 100,000 investment. *Capital.* https://www.capital.bg/biznes/kompanii/2018/11/28/3353236_chili_hils_fuds_razshiriava_proizvodstvoto_si_v_nov/   (in Bulgarian)

Hallett, S., Hoagland, L., & Toner, E. (2017). Urban agriculture: Environmental, economic, and social perspectives. *Horticultural Reviews, 44,* 65–120. https://doi.org/10.1002/9781119281269.ch2

Henriksen, K., Bjerre, M., Almasi, A., & Damgaard-Grann, E. (2012). *Green business model innovation – Conceptualization report.* Nordic Innovation Publication.

Hunold, C., Sorunmu, Y., Lindy, R., Spatari, S., & Gurian, P. L. (2017). Is urban agriculture financially sustainable? An exploratory study of small-scale market farming in Philadelphia, Pennsylvania. *Journal of Agriculture, Food Systems, and Community Development, 7*(2), 51–67. https://doi.org/10.5304/jafscd.2017.072.012

ISSK. (2020). *Urban agriculture as a strategy for improving the quality of life of urban communities.* Project website. Retrieved June 12, 2021, from https://www.urbanagriculture-bg.com/home

Jamal, M., & Mortez, S. (2014). The effect of urban agriculture in urban sustainable development and its techniques: A case study in Iran. *International Journal of Agriculture and Forestry, 4*(4), 275–285. https://doi.org/10.5923/j.ijaf.20140404.03

Kinkese, T., & Pride, C. (2017). The social, economic and health impacts of urban agriculture in Zambia. *Asian Journal of Advances in Agricultural Research, 3*(1), 1–8. https://doi.org/10.9734/AJAAR/2017/36720

Lohrberg, F., Lička, L., Scazzosi, L., & Timpe, A. (Eds.). (2016). *Urban agriculture Europe.* Jovis.

Nugent, R. (2003). Economic impact of Urban and Periurban agriculture. In ETC-RUAF (Ed.), *Annotated bibliography on Urban and Periurban agriculture* (pp. 1331–1371). SIDA, RUAF.

Petrova, V. (2014, April 19). *Food as a circumstance, food as a situation.* Seminar. Retrieved June 12, 2021, from https://www.seminar-bg.eu/spisanie-seminar-bg/broy10a/item/409-hranata-kato-obstoyatelstvo-hranata-kato-situaciya.html (In Bulgarian).

Pölling, B., Prados, M.-J., Torquati, B. M., Giacch, G., Recasens, X., Paffarini, C., Alfranca, O., & Lorleberg, W. (2017). Business models in urban farming: A comparative analysis of case studies from Spain, Italy and Germany. *Moravian geographical reports, 25*(3), 166–180. https://doi.org/10.1515/mgr-2017-0015

Simatele, D., & Binns, T. (2008). Motivation and marginalization in African urban agriculture: The case of Lusaka, Zambia. *Urban Forum, 19*(1), 1–21. https://doi.org/10.1007/s12132-008-9021-1

Simatele, D., Binns, T., & Simatele, M. (2012). Sustaining livelihoods under a changing climate: The case of urban agriculture in Lusaka, Zambia. *Journal of Environmental Planning and Management, 55*(9), 1175–1191. https://doi.org/10.1080/09640568.2011.637688

Smit, J., & Nasr, J. (1992). Urban agriculture for sustainable cities: Using wastes and idle land and water bodies as resources. *Environment and Urbanization, 4*(2), 141–152.

Van der Schans, J. W. (2010). Urban agriculture in the Netherlands. *Urban Agriculture Magazine, 24*, 40–42. Retrieved June 12, 2021, from https://edepot.wur.nl/411126

Van der Schans, J.W., et al. (2016). It is a business! business models in urban agriculture. In F. Lohrberg, L. Lička, & A. Timpe (Eds.), *Urban agriculture Europe* (pp. 82–191). Jovis.

Zasada, I. (2011). Multifunctional peri-urban agriculture – A review of societal demand and the provision of goods and services by farming. *Land Use Policy, 28*(4), 639–648. https://doi.org/10.1016/j.landusepol.2011.01.008

Chapter 6
Environmental Dimensions

Petko Tzvetkov, Ivaylo Dedov, Stoyan Beshkov, and Petar Shurulinkov

Abstract The environmental study within the project "Urban agriculture as a strategy for improving the quality of life of urban communities" took first steps to study some aspects of urban agriculture impact on environmental components of the City of Sofia. The quality of water used for irrigation, the content of heavy metals in soils and local produce were tested in the largest community garden in Sofia. Initial observations covered not only the positive impact but some of the main risks that should be considered while planning and developing UA in order to ensure human health and environmental protection. A brief literature review of the UA impact on other environmental factors was also made. Due to the limited project resources and time frame, we chose to put the focus of the study on the impact of UA on biodiversity. The contemporary character of urban fauna in Sofia is a result of the city's millennial history, during which people have cultivated and changed the environment. An initial assessment of UA impact on selected groups of organisms (land snails, butterflies, moths and birds) was made. The main processes that took place in the local fauna are: (1) degradation of the autochthonous fauna, and (2) introduction of allochthonous species and replacement of natural inhabitants. UA in cities has primarily social functions, while its role in maintaining biodiversity is rather secondary. A basis for further comprehensive research on the overall impact of UA and its contribution to sustainability of the urban environment and life was created.

P. Tzvetkov (✉)
Bulgarian Biodiversity Foundation, Sofia, Bulgaria

I. Dedov
Institute of Biodiversity and Ecosystem Research, Bulgarian Academy of Sciences, Sofia, Bulgaria

S. Beshkov · P. Shurulinkov
National Museum of Natural History, Bulgarian Academy of Sciences, Sofia, Bulgaria

© The Author(s), under exclusive license to Springer Nature Switzerland AG 2022
D. Pickard (ed.), *Urban Agriculture for Improving the Quality of Life*, Urban Agriculture, https://doi.org/10.1007/978-3-030-94743-9_6

6.1 The Impact of Urban Agriculture on the Environmental Components in the City of Sofia

6.1.1 Purpose of the Study

This study within the research project "Urban agriculture as a strategy for improving the quality of life of urban communities", described in more detail in Chap. 3, is designed to gather information about the impact of urban agriculture (UA) on the state of the environment (water, air, soil and biodiversity) as well as identify factors that affect its condition such as temperature, light, water retention capacity, carbon storage, the presence of specific pollutants in soil and in vegetables, waste management, including organic waste. In approaching this task, we were guided by literature that states UA benefits for the cities include "provision of high levels of biodiversity and ecosystem services that contribute to urban nature and environmental processes as well as a range of social benefits, such as food and nutrition, cultural resources and recreational benefits." (Lin et al., 2017).

6.1.2 Special Emphasis on the UA Impact on Biodiversity

Due to the limited project resources and time frame, a special focus was given to the impact of urban agriculture on biodiversity. Urban agriculture can provide and maintain high urban biodiversity and ecosystem services, including preservation of varieties and the diversity of agricultural crops and provision of ecosystem services such as pollination, resilience and protection against soil erosion, recycling of municipal waste, etc. (Lin et al., 2017). UA can provide diverse opportunities for observation and learning about the rich biodiversity, which involves the development of the so-called citizen science. An initial assessment of the impact of UA on biodiversity for three community gardens in Sofia was carried out through observations on selected groups of animals. The results of this assessment are presented in the Sect. 6.2 of this chapter.

6.1.3 Potential UA Impact on Urban Environmental Problems in Sofia

When reviewing urban environmental problems, which can be influenced to some extent by UA, the following topics stand out in the case of Sofia City: reducing the effect of the urban heat island, providing quality water and protecting water sources from pollution or the destructive effects of runoff water, protecting and improving soil quality, conservation of biodiversity and ecosystem services, reducing the carbon footprint, the potential contribution to improving the condition of polluted urban areas, stimulating recycling and sustainable management of organic household waste through composting, efficient use of urban and suburban areas, commitment to environmental issues by urban communities.

Most of these claims can only be confirmed in a larger study in an environment where UA is widespread. Since known community gardens for UA in Sofia City cover just about 1.6 acres in August 2020, they should be analysed as a part of the larger city's green system. Subsequent large-scale research should be part of a comprehensive study of the ecological role of the green system on the environmental components in the capital city. This preliminary study does not cover the private gardens within the urban and peri-urban areas, but only community gardens (Figs. 6.1 and 6.2).

Due to the relatively small area of the community UA gardens in the City of Sofia and due to their location in an urban environment that varies in terms of urbanization degree and in view of the project's limited resources, it was not possible to carry out in-depth studies on the impact of urban agriculture on abiotic components and environmental factors. Under this project at the largest site – ZaDruzhba community garden – we sampled and analysed in a once-off exercise: (1) the water quality in the sole water source – a well, and (2) soil for content of heavy metals and samples for the presence of heavy metals in vegetables grown at the model site. This was done in order to provide reliable data on the suitability of this garden to provide safe soil and water for food production and to test popular beliefs that it is harmful to consume the products of UA because of the polluted urban environment (see Chap. 2).

Fig. 6.1 Community garden in Druzhba area in Sofia (ZaDruzhba garden)

Fig. 6.2 Community garden in Studentski grad area in Sofia

6.1.4 Irrigation Water Quality in ZaDruzhba Community Garden

To test water quality, we took water samples on 27.01.2020 from the well, with coordinates N 42.644337° E 23.417824°, which was tested at the Central Laboratory Sofia 01 of the Environment Executive Agency (EEA) of the Ministry of Environment and Water (MoEW). The test covered the following water parameters: nitrates, nitrites, ammonium, total hardness (the concentration of calcium and magnesium), suspended solids, total alkalinity, carbonates, hydrogen carbonates, sodium, potassium, calcium, magnesium, chlorides, sulphates, orthophosphates (such as PO4), total iron, manganese, lead, cadmium, mercury, gross beta activity.

The test results, as per the report of the Central Laboratory Sofia 01 of the EEA of the MoEW show that all tested indicators are within the norm and do not exceed the requirements for water quality for irrigation set out in Ordinance No 18 from May 27, 2009 on the quality of water for irrigation of agricultural crops, issued by the Minister of Environment and Water and the Minister of Agriculture and Food. On the basis of the report, the following recommendations for further research, monitoring and conservation of water quality were made:

- To monitor on a regular basis the quality of water in surface or groundwater sources used for irrigation. To ensure periodic monitoring of the water quality in connected or potentially affected water bodies from the UA areas.
- To examine and subsequently monitor all indicators of irrigation water outside the scope of the initial test, including sanitary and hygiene indicators, petroleum products and pesticides.

Fig. 6.3 Buffer water container for irrigation in ZaDruzhba communal garden

- To avoid artificial fertilizers and pesticides in UA, due to possible lasting consequences for the health of citizens and permanent pollution of water sources.
- Not to use water that does not meet the quality criteria for irrigation water set out in regulations in force (Fig. 6.3).

6.1.5   Quality of Soil and Vegetables in ZaDruzhba Community Garden

The soil in the urban environment is in most cases compacted and fragmented. In ZaDruzhba community garden, the soil is alluvial and falls in the alluvial terrace of the Iskar River (Tsolova & Tomov, 2018). A number of methods and practices, such as cover cropping, mulching, producing in raised beds, and changing subsurface drainage through piping, can improve soil conditions to support food production (Beniston & Lal, 2012, Lin et al., 2017). In addition, the application of compost matter retains and restores the soil organic matter content. Still, there is skepticism in the public opinion about how suitable the soil in this garden is for food production as it is located in the east of the city and pollution with heavy metals from the socialist times is assumed (see Chap. 2).

To test the concentration of heavy metals in the soil and produce in the ZaDruzhba community garden, in 2019 we sampled soil from the site under consideration, as well as two pairs of identical vegetables, carrots and kale: one sample produced in the garden and one bought from a supermarket chain.

The soil analysis was governed by the requirements of Ordinance No 3 from August 1, 2008 on the norms for allowed content of harmful substances in soils. Heavy metals such as total content in mg/kg dry soil were analysed by aqua regia extraction under a standardized procedure (ISO 11466). The following norm-bound heavy metals were detected by atomic absorption spectrophotometry of the Perkin Elmer 2100 device: copper, zinc, cobalt, nickel, lead. Cadmium, arsenic and mercury were not detected out of the norm-bound elements, but elements outside the scope of the norm setting such as iron and manganese were detected.

Ordinance No 3 on the norms for allowed content of harmful substances in soils gives the following definitions: "Precautionary concentration" (PC) is the content of harmful substances in soil in mg/kg, whose exceedance does not lead to disturbance of of soil functions and danger to the environment and human health.

"Maximum allowable concentration" (MAC) is the content of a harmful substance in soil in mg/kg which, if exceeded, can lead to disturbance of soil functions and to danger for the environment and human health.

The results of the tests were analysed for the project by Prof. Mariana Hristova from the Institute of Soil Science, Agrotechnologies and Plant Protection at the Bulgarian Agrarian Academy. Figure 6.4 draws a comparison based on the soil analysis for heavy metals benchmarked to PC and MAC of the elements as per Ordinance No 3.

These results show that the content of copper and zinc is higher than PC, but is far below the MAC. The soil has neutral to slightly alkaline pH. The contents of elements cobalt, nickel, chromium and lead are lower than the PC. Based on the soil analysis for heavy metals, we can say that the soil does not contain above-MAC heavy metals and hence it is not harmful to human health and can be used to grow edible plants.

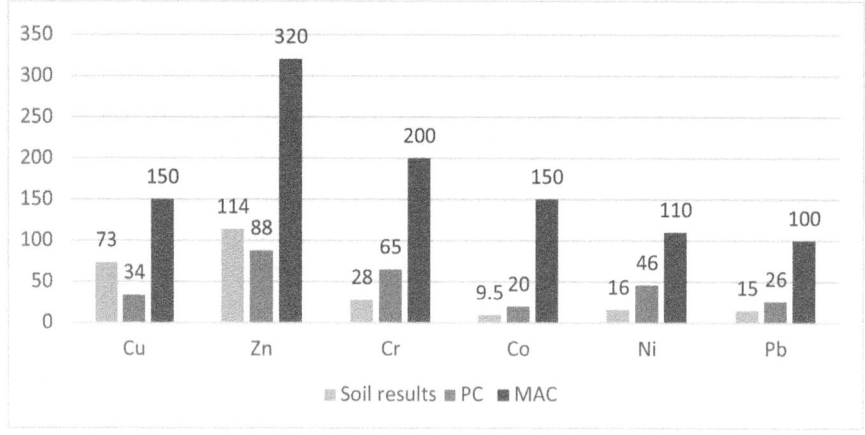

Fig. 6.4 Comparison of the content of heavy metals (mg/kg) in the soil of ZaDruzhba garden to PC and MAC values regulated by Ordinance No 3

The sampling of vegetables grown in ZaDruzhba garden for analysis in comparison with the respective products purchased from the supermarket chain is a good approach to testing whether the consumption of vegetables from the garden is safe for human health. Out of the norm-bound elements, we found lead and cadmium in the plants (kale and carrots) grown in the garden and the retail network. Their MACs are set in accordance with Regulation (EC) 1881/2006 and detailed in Ordinance No 5 from 2015 setting maximum levels for certain contaminants in foodstuffs. These documents set the MAC for cadmium at <0.100 mg/kg and for lead at <0.30 mg/kg (wet weight).

Comparative measurements for the content of lead, cadmium, zinc and nickel in vegetables from the garden and those from the retailer (Fig. 6.5): show 1. No concentrations of lead and cadmium were found in the plants grown in ZaDruzhba garden. 2. The zinc content is higher in kale grown in ZaDruzhba garden, but it is not dangerous due to its safe content in the soil. 3. The zinc content in carrots was found to be lower in plants grown in ZaDruzhba garden. 4. The nickel content is twice lower in carrots and kale grown in the garden, which is understandable because the nickel content in the soil is below the national average.

The conclusion from these tests is that no concentrations of harmful substances for copper, lead, zinc, nickel, cobalt and chromium have been detected in the soil, which if present could lead to disturbance of soil functions and to danger to the environment and human health according to Ordinance No 3 on the norms for allowed content of harmful substances in soils. No concentrations of lead or cadmium have been detected in the plants as per Ordinance No 5 of 2015 setting the maximum levels of certain contaminants in foodstuffs.

The soil and water quality tests in Sofia's largest urban agriculture garden and a comparative analysis of the levels of heavy metals in vegetables grown in this garden

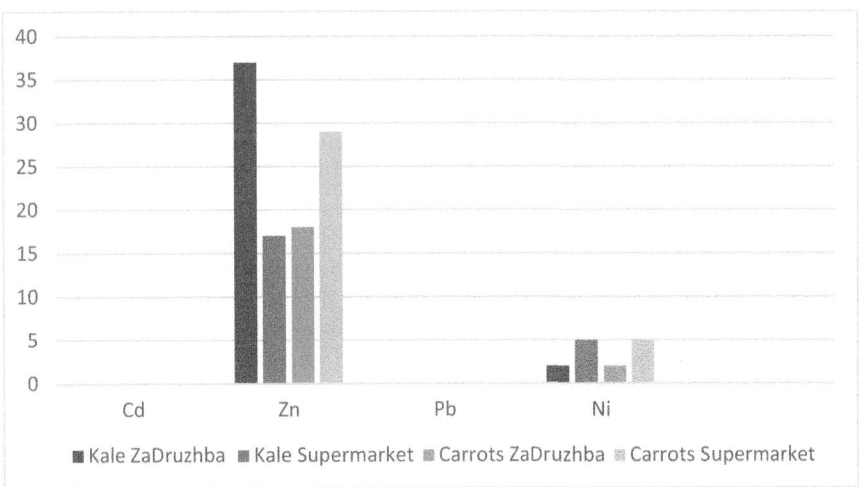

Fig. 6.5 Comparison of the contents of heavy metals (mg/kg) in kale and carrots grown in ZaDruzhba with the same type of vegetables from the supermarket chain

and those purchased from a large retailer prove that at least in regards these elements, it is possible to grow clean food in cities. Still, additional testing for harmful substances is recommended in all collective gardens, including analysis of polycyclic aromatic hydrocarbons (PAHs) which are more common in dense urban areas and along busy traffic routes.

Recommendations for UA practitioners and policy makers:

1. Further tests should cover the full range of harmful substances in soil, i.e. to further study the content of cadmium, mercury and arsenic (Ordinance No 3/2008). For plants, norm-bound elements to assess harmful concentration are: lead, cadmium, arsenic and mercury (Ordinance No 5/2015). And the measurements should be performed in an accredited laboratory.
2. Also, to conduct tests on other potential soil contaminants such as petroleum products, pesticides, etc.
3. Other studies may include and be related to the study of organic matter content and carbon storage capacity.
4. Fertilizers and pesticides should be avoided in UA.

6.1.6   Other Impacts

The community gardens across Sofia make use of composting household organic waste and its application in fertilization of plots, which aims to close the carbon cycle and maintain soil quality by preserving the organic matter content in soil. Composting is applied in all three community gardens in Sofia. In addition to the reduction of organic waste, UA and sharing produce are shortening the food supply chain and thus avoiding unnecessary packaging.

There is great potential of UA to reduce the urban heat island effect due to the increased evapotranspiration from vegetation and water bodies (Qiu et al., 2013). There is a positive example of creating a community garden for UA in inter-block spaces to replace previous parking lots in community gardens in Sofia Studentski grad district, which changes the ground surface from concrete to soil and vegetation (Fig. 6.2). Also, artificial ponds have been created in two of the three community gardens in Sofia (Fig. 6.6).

Furthermore, UA gardens provide access and contact with the natural environment, even in high-rise residential buildings where some UA practitioners grow herbs, vegetables and flowers, including on their balconies (Fig. 6.7).

Involvement of local communities also provides a better understanding of and support for addressing environmental issues and shaping attitudes towards the environment and food. UA is also seen as a potential measure to support the food needs of growing urban populations and can contribute to further development of the perceptions of "local and seasonal food" as well as "healthy and sustainable food" and thus to promote and develop responsible consumption patterns and attitudes towards the environment (Feola et al., 2020). This includes socially significant practices such as providing food for disadvantaged groups, which is a tradition

Fig. 6.6 Artificial pond in ZaDruzhba garden

Fig. 6.7 Urban gardening on the balcony of a private home

established in the ZaDruzhba community garden. The close relation of environmental and social impacts of UA is also demonstrated by the interest in UA that many environmental organizations and civil groups share. These include not only the practitioners in the existing community gardens in Sofia, but also active interest groups and civil organizations that work at the intersection of UA development and sustainable urban development and environmental protection, such as groups encouraging composting, balcony food gardening, etc.

Through these impacts, UA further enhances the positive social effects of UA such as supporting the urban farming community with new knowledge and skills, and the choice of healthy and self-produced food has a positive impact on the quality of life in the city.

In the frame of this project the first steps were undertaken to study some aspects of UA impact on the environmental components of the City of Sofia. Issues related to water quality used for irrigation, the content of heavy metals in soils and their presence in local produce were addressed. In essence, these initial observations address not only the positive impact but some of the main risks that should be taken into account while planning, developing and practicing urban agriculture in order to ensure human health and environmental protection. A brief literature review of other aspects of the UA impact on other environmental factors was also made. The UA effects on certain selected groups of animal organisms are considered in a separate chapter. All this is a good basis for further comprehensive research on the overall impact of urban agriculture, not only on the environment and communities, but as a whole its contribution to the sustainability of the urban environment and life.

6.2 Impact of UA Spaces and Practices on Biodiversity. Initial Data Based on Model Groups

6.2.1 Formation and Nature of Urban Zoocenoses[1] in Sofia

The contemporary character of urban fauna in Sofia is a result of the city's millennial history, during which people have cultivated and changed the environment to meet their own needs. It can be assumed that in ancient times the human influence on biocenoses[2] was weak, but as the technical evolution of mankind proceeded, this influence became more and more noticeable and the communities of plant and animal species moved further and further away from their original nature. The main processes that took place in the local fauna can be summarized in two points:

1. degradation of the autochthonous[3] fauna – a process running parallel to the formation and change of the landscape, and

---

[1] An interconnected set of all animals inhabiting a distinct part of the environment, be it water or land.

[2] The totality of all living beings inhabiting a given territory/space.

[3] Local, primary, originated in this place.

2. introduction of allochthonous species[4], respectively to a larger or smaller extent (depending on the group) replacement of the natural inhabitants of the area.

The above two processes are characterized by the loss of rare and endemic species and the invasion of widespread species, over-dominance and eventually the formation of typical urban cenoses. Urban cenoses, on the other hand, in different parts of Europe, manifest more similarities with each other than a city with its surroundings.

For example, the land snail fauna in the City of Sofia, while composed of various malacocenoses[5], shares some common features that distinguish it from the fauna of the Sofia surroundings. In contrast to Sofia surroundings, the Sofia malacofauna[6] demonstrates changes in the zoological structure – the share of a local species decreases (Bulgarian endemics, Balkan endemics and European mountain species) and the share of invasive species increases (Mediterranean, European and Holarctic). Also, in Sofia, although slightly, the share of drought-tolerant species and eurybionts[7] is growing, at the expense of declining mesophilic[8] and moisture-loving species (meso-hygrophiles and hygrophiles). At present, the Sofia urban malacofauna has features of anthropogenic snail fauna from the southern parts of the continent – a larger share of Mediterranean species, respectively, a higher share of drought-resistant and heat-loving species, as well as a smaller number of species in general, compared to the more northern urban fauna.

When it comes to Lepidoptera, some authors are of the opinion that urban communities are separate, have a distinct character and are different from non-urban ones (the Noctuidae family of Lepidoptera – Winiarska, 1986). In urban environments, one or two species of a group are most numerous and accordingly the index of dominance increases (Winiarska, 1986).

Birds also represent a significantly poorer urban avifauna compared to suburban areas. Many rare species are absent in the city, despite the presence of agricultural gardens or well-forested and relatively sparsely visited parks. In peri-urban areas there is still much less disturbance, and habitats are quite diverse. Some examples of species present in the Sofia area and nearby foothills, which are not nesting birds in the city itself, are the common cuckoo, Eurasian hoopoe, European bee-eater, willow tit, sombre tit, Eurasian skylark, etc. There are also large groups of waterfowl and diurnal birds of prey, of which only a small number of the species enters the City of Sofia to nest, but are found in its vicinity.

The unification of urban zoocenoses into local urban fauna is underpinned by some differences in the typological parameters of communities in the city. Namely the parameters that distinguish the city from its surroundings shape the character of the urban fauna. These are: species diversity, abundance, the number/species ratio

---

[4] Imported, of non-native origin.

[5] The molluscs that inhabit an area.

[6] The mollusc species complex that inhabit a given area.

[7] Plant or animal species that tolerate widely changing environmental conditions.

[8] An organism that lives under moderate environmental conditions.

and the ecological structure of communities. Similar to all urban fauna, the differentiation of urban zoocenoses (malacocenoses) is due to the presence (in many cases also high abundance) of some allochthonous species. These species, present only (or mainly) in urban environments, can be called "indicators" for urban zoocenoses. It is these allochthonous species that precondition the high species diversity in urban biotopes, in contrast to non-urban ones, where the high diversity is a result of the presence and high abundance of autochthonous stenobiotic species.

6.2.2 Purpose, Model Groups of Organisms and Territorial Scope of Research

6.2.2.1 Purpose of the Study

The aim of this study was to make an initial assessment of the impact of urban agriculture on selected groups of organisms on the territory of Sofia City.

6.2.2.2 Model Groups of Organisms

To enable the objective assessment of the state of the environment, and an adequate forecast for its development, it is necessary to either use a set of organisms, or the group we use must meet certain conditions. The relevant group of organisms needs to be widespread, sufficiently numerous as species and specimens, to occupy a wide range of diverse biotopes and to react to changes in the environment.

Land snails not only meet these requirements but also have some additional advantages: slow mobility, which makes them sensitive to changes in living conditions; relatively high biomass, which allows easy analysis of their tissues in view of heavy metal contamination; various environmental preferences (from stenobionts to eurybionts); various drought resistance and moisture resistance (from xero-, meso-, hygro- to hydrophiles); various trophic preferences (herbivores, omnivores, predators). All this makes land snails a desirable object of research on the state of the environment and monitoring of processes in it (Gomot de Vaufleury & Pihan, 2000).

Butterflies, moths and birds also largely meet these requirements, and in particular in view of their dependence on and preference for the plant species grown in urban agriculture that serve as food or for reproduction. Furthermore, their attractiveness and people's interest in butterflies and birds are a good starting point to involve UA practitioners in further studies on these groups of organisms.

The assessment of land snails was carried out within several observations and soil sampling in the summer of 2020.

To assess butterflies and moths, we made two observations in the autumn (August and September) of 2019, covering the butterfly superfamily of *Papilionoidea,* and diurnal representatives of the *Crambidae* family, spotted flying in daytime.

For birds, observations took place repeatedly in the 1990–2010 period (personal data of the author) and in the autumn of 2019. The observations were carried out in all seasons and annually. A series of catches with bird nets were also conducted in order to determine the species composition and ringing. They were carried out mainly during the spring and autumn migration of birds.

6.2.2.3 Territorial Scope

For the purposes of the study, we selected the three community gardens for UA, which have been cultivated in recent years and are available for observation. Also, due to time and financial constraints, the observations were carried out only on gardens located in or in close proximity to the actual urban/highly urbanized part.

(A) *Community garden for urban agriculture in Druzhba 2 residential district (ZaDruzhba garden)*

Coordinates (at the entrance): N42.64405 E023.41771, 560 m above sea level.

The garden is located near the Iskar River and its adjacent wetlands. It borders on natural woods, bushes and open terrains. The garden is positioned in a fenced space with a greenhouse and a site for intensive cultivation of flowering plants. A small artificially maintained reservoir has been created amid the arable land, which apparently serves to attract insects and develop aquatic vegetation. A drill/well is used for irrigation. The garden features: pumpkins, tomatoes, cucumbers, peppers, beans, cabbage, onions, garlic, maize, sunflower, parsley, dill, beetroot, carrots, okra, Jerusalem artichoke (earth apple), raspberries, strawberries, asparagus. There are single greengage and apple trees. Honey and essential oil plants are represented by mint, oregano, rosemary, savory, basil, lavender, spearmint. The flowers grown are marigold, rose, white water rose, tagetes, petunia, etc (Fig. 6.8).

(B) *Community garden for UA in Studentski Grad*

Coordinates (at the entrance): N42.65496 E023.33848, 602 m above sea level.

The garden is in an inter-block space with access limited by a mesh fence. In the immediate vicinity to the south and east there are single greengage trees. To the north across the road, there is an area heavily overgrown with walnut, hazel trees and blackberry with flower greenhouses. The garden is irrigated artificially. The plants grown in the garden include: pumpkins, watermelon, tomatoes, cucumbers, peppers, beans, leeks, corn, sunflowers, parsley, dill, carrots. Mint, basil and spearmint represent honey and essential oil plants. Flowers include geranium, carnations, tagetes (wild marigold), decorative vine (Fig. 6.9).

(C) *Community garden for UA in the yard of the Knyaginya Evdokia Vocational High School for Hairdressing and Cosmetics*

Coordinates (at the entrance): N 42.673056 E023.355833, 571 m above sea level.

Fig. 6.8 Views from ZaDruzhba garden

The garden is in a relatively shady area in the school yard and inter-block space, and the access to it is limited. There is artificial irrigation and a small artificial pond that temporarily retains water. The garden crops are: pumpkins, tomatoes, cucumbers, peppers, maize, sunflowers, potatoes, strawberries, cabbage. In the immediate vicinity there are poplar trees. Honey, essential oil plants and flowers are not grown (Fig. 6.10).

Fig. 6.9 View from community garden in Studentski grad

6.2.3 Preliminary Results in the Study of the Model Groups in the UA Spaces

(A) *Land snails*

The snapshot of land snail populations in areas used for agriculture within the city of Sofia shows that these are urban malacocoenoses, but with some specificities. The snail fauna is relatively rich (a number of taxa were found only as juvenile shells and cannot be identified by species – *Cecilioides* sp., Clausiliidae, *Monacha* sp., *Truncatellina* sp.), but the summer period, during which the samples were collected, gives only a partial idea of the species composition. These are typical urban malacocoenoses, and relatively more common species are present (Caucasotachea vindobonensis (C. Pfeiffer, 1828), Chondrula tridens (O. F. Müller, 1774), Deroceras cf. turcicum (Simroth, 1894), Helix lucorum Linnaeus, 1758, Punctum pygmaeum (Draparnaud, 1801), Vertigo pygmaea (Draparnaud, 1801), Zebrina detrita (Müller, 1774) and there are no more local or endemic taxa. Two invasive species were identified (Arion) cf. lusitanicus (J.Mabille, 1868), Deroceras cf. sturanyi (Simroth, 1894), along with a large percentage of xero-resistant and meso-xero resistant. Despite the lack of quantitative studies, the dominance of eurybiotic *H. lucorum* is obvious. The landscape of the studied habitats (open arable gardens, surrounded by xerophilous meadows and single trees and shrubs) explains the high percentage of heat-loving and drought-tolerant species – xero-resistant, meso-xerophilous and eurybiont. Shrubs and denser shadows under the trees around the gardens are the cause of the presence of mesophilic and/or forest species such as *D. stutanyi, D. turcicum, P. pygmaeum*. The presence of these species, as well as the

Fig. 6.10 Community garden for UA in the yard of the Knyaginya Evdokia Vocational High School for Hairdressing and Cosmetics

relatively more hygrophytic *A. lusitanicus* and *V. pygmaea,* is further aided by the presence of an irrigation system that provides much-needed moisture to snails even in drier periods. It is assumed that the invasive organism *A. lusitanicus* uses shady and relatively wetter shelters, from where, in wetter weather, it makes food

Fig. 6.11 Helix lucorum in ZaDruzhba community garden

migrations to the beds. Although not as aggressive *H. lucorum* can also be a threat to cultivated plants due to its high numbers (Fig. 6.11).

Somewhat surprising is the presence of the subterranean genus *Cecilioides* in the Druzhba garden. One possible explanation is the proximity of the Iskar River, surrounded by suitable habitats for the species, which is also a possible corridor for the transmission of microgastropods.

The main conclusion to be drawn at this stage is that arable gardens within the cities maintain typical urban malacocoenoses, as the irrigation systems and the available food resources further stimulate the presence and abundance of some species.

(B) *Butterflies and moths*

The butterfly and moths fauna is poor – 11 species were found, but this is typical of this period of the year (August – September 2019). The following species were found: Carcharodus alceae (Esper, [1780]), Pieris rapae (Linnaeus, 1758), Colias crocea (Geoffroy in Fourcroy, 1785), Lycaena dispar (Haworth, 1802), Leptotes pyrithous (Linnaeus, 1767), Polyommatus icarus (Rottemburg, 1775), Cynthia cardui (Linnaeus, 1758), Maniola jurtina (Linnaeus, 1758), Coenonympha pamphilus (Linnaeus, 1758), Nomophila noctuella ([Denis & Schiffermüller], 1775), Pyrausta sanguinalis (Linnaeus, 1767). The detected representatives are almost certainly accidentally caught, i.e. they have flown as an imago[9], rather than being fed and metamprphosed into adult form here. Their accidental emergence is a consequence of the presence of honey and essential oil plants such as mint, oregano,

---

[9]The last phase of insect metamorphosis, when the individual has reached sexual maturity.

rosemary, savory, basil, lavender, on which adult butterflies land and feed. In addition to butterflies and moths, bees and two-winged insects have been spotted landing and feeding on the flowers of these plants. Cultivated plants, even during flowering, are less attractive to insects than honey and essential oil plants. The blooming flower vegetation attracts passing insects and insects from the surrounding areas, and they fly in to feed. This is especially noticeable during the dry summer months, when the surrounding vegetation is already in post-bloom and dry, and that in the cultivated gardens is fresh and in bloom as a result of irrigation care. In fact, wastelands, uncultivated plots and insufficiently maintained city parks have a greater diversity of species. Out of the studied UA gardens, the one in Druzhba manifests the richest range of species; it has the largest area, located on the edge of the city and is surrounded by undeveloped heterogeneous and relatively well-preserved natural habitats (Figs. 6.12 and 6.13).

If observed during the months of May-July, when most butterfly species breed and lay eggs and when cultivated plants are in bloom, the species diversity is expected to be significantly higher. Biodiversity could be enhanced if the area of community gardens for UA is increased and there are corridors connecting them with parks, meadows and wastelands. The isolated small gardens attract for a short time only the passing butterflies and moths, and this is attributed to the flowers there.

(C) *Birds*

The avifauna of urban gardens used for agriculture in Sofia is characterized by a great degree of naturalness and consists mainly of autochthonous species. Of course,

Fig. 6.12 Leptotes pirithous in ZaDruzhba community garden

Fig. 6.13 The blooming honey and essential oil plants such as lavender are attractive for butterflies

the percentage of synanthropic[10] species is high, but they are also native. This is a major difference compared to large cities in Western and Central Europe where the number of alien and invasive bird species is too high. The degree of urbanization of birds in our country is generally lower than in other parts of Europe and many species that nest widely in similar gardens and parks there; in our country the following are not yet synanthropic or are in the initial stage of synanthropization – the common wood Columba palumbus (Linnaeus, 1758), stock dove Columba oenas (Linnaeus, 1758), etc. However, certain species that are rare and scarce elsewhere in Europe are numerous synanthropic species in Sofia – such as pallid swift Apus pallidus (Shelley, 1870) and Alpine swift Tachymarptis melba (Linnaeus, 1758).

In general, the avifauna of ZaDruzhba garden and its surroundings is quite rich, represented not only by typical synanthropic species but also by many species typical of forest and bush habitats. The other gardens for UA are inhabited mainly by typical synanthropic species, especially during the nesting period – semi-domestic pigeons Columba livia (J. F. Gmelin, 1789), house sparrow Passer domesticus (Linnaeus, 1758), starlings Sturnus vulgaris (Linnaeus, 1758), etc. Outside the nesting period, visits of species originating from city parks or from areas adjacent to the city, such as blackbird Turdus merula (Linnaeus, 1758), song

---

[10] An organism whose existence is linked to human activity.

thrush Turdus philomelos (C. L. Brehm, 1831), tree sparrow Passer montanus (Linnaeus, 1758, great tit Parus major (Linnaeus, 1758), blue tit Cyanistes caeruleus (Linnaeus, 1758), common chaffinch Fringilla coelebs (Linnaeus, 1758), hawfinch Coccothraustes coccothraustes (Linnaeus, 1758), greenfinch Chloris chloris (Linnaeus, 1758), etc. are more common.

The richness of the avifauna at ZaDruzhba garden is largely due to the great number of transit migrating birds and the border position of the place in the outer city. The thorough research of the area around the garden also provides substantial information, as a result of the author's own research and data from the period 1990–2010. The nesting avifauna of the area around and in this garden includes 66 species, which is not a small number for such an insignificant space. The number of detected species detected is 137, 119 of them arenprotected. The presence of natural or semi-natural habitats in the vicinity of the garden is, in the case of this garden, also of key importance for the species richness. These are available in this suburb part of Druzhba, but are absent in some of the other similar gardens in Sofia (Fig. 6.14).

In the community gardens for UA around Sofia there are mainly from the orders of Passeriformes and Columbiformes. This is due to the food resources available in these gardens, which attract certain bird species, mainly from these orders. Representatives of the other orders enter these places far less frequently and more or less by chance, mainly during migration and wintering.

In terms of zoogeography, the avifauna of the City of Sofia and the UA gardens does not differ significantly from the avifauna of the lower parts of the Sofia region – with species of European, Palearctic, Holarctic origins and species of the fauna of the Old World being predominant. The Mediterranean, Turkestan-Mediterranean, steppe and boreal species have a low prevalence.

The above is reflected in the fact that for biodiversity in all three model groups, the surrounding habitats are more important – shrubs and groves (in the present study we refer to the garden near the Iskar River) – than the arable areas. It is these preserved green spots around and in the terrains used for urban agriculture that turn out to be oases for many species, which enter and feed in the gardens under suitable weather conditions.

6.3   Lessons Learned

Historically, the change of the natural environment in agricultural plots, orchards or vegetable gardens, etc., can be seen as an intermediate stage in the transformation of natural communities into urban or vice versa in the modern context. In this sense, the practice of urban agriculture is a "step in the right direction" to preserve the connection of man with nature and relieve stress from urban dynamics.

In terms of biodiversity conservation, the initial assessment appears to demonstrate that agricultural areas are of less importance. On the one hand, these are places

Fig. 6.14 Eurasian jay Garrulus glandarius and common chaffinch Fringilla coelebs are common visitors of community gardens in Sofia

rich in food resources – a fact that is important for all three model groups – snails, butterflies and moths, and birds. Also, the regime of regular irrigation favours many organisms and this is especially evident in land snails. On the other hand, these areas can be subjected to frequent mechanical (ploughing, weeding, mowing, cleaning, etc.) and chemical (fertilization, treatment with pesticides) treatment. In addition, these, in most cases, open and unprotected from direct sunlight areas (treated, tilled land, not covered by vegetation, including woodland), further complicate the survival of some species of land snails.

With regard to butterflies and moths, the blooming flower vegetation is a key factor of species diversity. It attracts passing insects and insects from the surrounding areas and they fly in to feed. It is recommended:

- To plant and grow honey and essential oil plants such as mint, oregano, rosemary, savory, basil, lavender by and amid vegetables and along fences and pathways. Apart from butterflies and moths, bees and two-winged insects have been spotted landing and feeding on the flowers of these plants. I also recommend planting and growing more and longer-lasting flowers. The blooming flower vegetation attracts insects from the surrounding terrains and they fly in to feed. This is especially noticeable during the dry summer months, when the surrounding vegetation is already in post-bloom and dry, and that in the cultivated gardens is fresh and in bloom as a result of irrigation care.
- To conduct similar observations in May-July, when most butterfly and moth species breed and lay eggs and the species diversity is significantly higher.
- For birds, a limiting factor is the small area of UA gardens. Without surrounding suitable habitats, they would not be able to maintain a rich avifauna on their own because these are too small – often not exceeding the nesting area of one or two pairs of passerines. As a result, there is a significantly poorer avifauna in the UA gardens within the city, surrounded by buildings and other bird-unfriendly habitats, compared to the garden in the outer parts of Druzhba. The diversity of birds in such a garden is determined by the following main factors:
- The presence of food in the garden itself. Flower plants attract insects which are food for many types of passerines. Also, plants, their seeds and fruits are an attractive food resource for some bird species. Fruits are especially attractive to some migratory birds in the autumn, before they fly away.
- The presence of shrubs and low-tree vegetation in the garden and along the margins. In the case of ZaDruzhba garden there are shrubs and low fruit trees that attract blackbirds, thrushes, European robins, woodpeckers, warblers, shrike, sparrows, starlings, etc.
- The presence of various habitats in the neighborhood – poplar, willow and alder groves, shrub formations, riparian vegetation, river, open habitats – meadows and arable land, urban habitats.
- The presence of a migration route above or near the garden. Such a route passes along the Iskar River during the spring and autumn migration and is part of the migration route Via Aristotelis.
- The presence of natural or semi-natural habitats in the vicinity of the garden. These are available in the outer part of Druzhba residential district, but absent in some of other similar gardens in Sofia. In the case of the Druzhba 2 garden these are riverside forest and brush habitats along the Iskar River. There are also small wetlands between the river and the garden – holes filled with water and covered with riparian vegetation. The nearby habitats include poplar, willow and alder groves, shrub formations, riparian vegetation, river, open habitats – meadows and arable land, urban habitats.

To make them more attractive for birds, when managing UA gardens, it is recommended:

- When choosing a place, these gardens should be close to the outer boundaries of the city or to a park, reservoir, or at least large green spaces.
- The garden itself should have fruit trees and shrubs – greengage, quince, medlar, cherry, pear, apple trees, and elderberry, currant, blackcurrant, dogwood, rowan, rose hip, raspberry, etc. The greater number of such trees and shrubs can attract some birds to nest in the garden itself.
- Out of the cultivated herbaceous plants, it is good to have honey and essential oil species, as well as flowers to attract more insects. This will further attract more birds.
- During the winter months, if possible, to organize the feeding of birds in the garden with seeds.

When implementing the recommendations, the urban community gardens can become attractive spots for many bird species, oases for birds amid a highly urbanized environment. This would undoubtedly be a very useful and environmentally friendly practice in this regard.

In connection with the monitoring and collection of additional data, it is recommended that people involved in community gardens should be trained to operate any of the mobile applications such as ObsMapp, Smart Birds, etc. or they may join thematic groups in social media, where they will also receive help to identify pictures of butterflies and moths, and birds, and the data will be used to continue the observations. If there is good will and resources for administration, it would be best to create a dedicated Facebook group for UA gardeners to share pictures of butterflies, birds, and other animals to identify and monitor the species.

In conclusion, we can infer that UA practices in cities primarily have social functions – to teach work ethic, contact with nature, to provide stress relief, to produce fruits and vegetables for households, etc., while their role in maintaining biodiversity is rather secondary.

References

Beniston, J., & Lal, R. (2012). Improving soil quality for urban agriculture in the North Central U.S. In R. Lal & B. Augustin (Eds.), *Carbon sequestration in urban ecosystems* (pp. 279–313). Springer.

De Vaufleury, A., & Pihan, F. (2000). Growing snails used as sentinels to evaluate terrestrial environment contamination by trace elements. *Chemosphere, 40*(3), 275–284. https://doi.org/10.1016/S0045-6535(99)00246-5

Feola, G., Sahakian, M., Binder, C., & Zundritsch, P. (2020). Sustainability assessment of urban agriculture. In C. Binder, R. Wyss, & E. Massaro (Eds.), *Sustainability assessment of urban systems* (pp. 417–437). Cambridge University Press. https://doi.org/10.1017/9781108574334.019

Lin, B. B., Philpott, S. M., Jha, S., & Liere, H. (2017). Urban agriculture as a productive green infrastructure for environmental and social well-being. In P. Tan & C. Jim (Eds.), *Greening cities: Forms and functions, advances in 21st century human settlements* (pp. 155–179). Springer. https://doi.org/10.1007/978-981-10-4113-6_8

Qiu, G., Li, H., Zhang, Q., Chen, W., Liang, X., & Li, X. (2013). Effects of evapotranspiration on mitigation of urban temperature by vegetation and urban agriculture. *Journal of Integrative Agriculture, 12*(8), 1307–1315. https://doi.org/10.1016/S2095-3119(13)60543-2

Tsolova, V., & Tomov, P. (2018). Morphological and classification hallmarks of soils in green zones of Sofia city. *Soil Science, Agrochemistry and Ecology, 52*(3), 43–56. (In Bulgarian).

Winiarska, G. (1986). Noctuid moth (Lepidoptera, Noctuidae) communities in urban parks of Warsaw. *Memorabilia Zoologica, 42,* 125–148.

Chapter 7
Spatial Dimensions

Nina Toleva-Nowak

Abstract The spatial study included desk research regarding various historical, theoretical and regulatory aspects, as well as a review and analysis of specialized literature, historical data, archive maps and drawings. This stage was followed by extensive and thorough field work, exploring the urban and architectural specifics of three very distinct types of urban environment – a classical high intensity intra-urban district, a peri-urban "bedroom community" district, and a suburban region including ten rural settlements and their adjacent territories. Several types of urban agriculture practices were traced and analysed and the data were mapped in their GPS coordinates and represented as interactive maps.

The results showed how urban agriculture practices differ depending on the specifics of the district. The produced statistics, graphics and maps proved that the centuries-old bottom-up traditions enhanced with contemporary socio-cultural practices are valid approaches for improving quality of life on the individual, communal and city level, mostly affecting the actual internal and the potential external dimensions.

7.1 Introduction and Short Historical Overview

Urban agriculture is a multifaceted phenomenon. As it occurs in the city fabric, its spatial aspect cannot be left aside. On the contrary, it needs deeper examination – not just historical and theoretical, but also normative, practical and in situ.

Although the mainstream interest in urban agriculture practices is believed to have begun in the late 1980s and early 1990s, its roots on the international scene can be traced back as early as the end of the nineteenth and the beginning of the twentieth century, when in 1898 Ebenezer Howard proposed the idea of garden cities. His urban concept was based on strictly planned singular communities, framed by green belts in order to ensure optimal balance between the residential, industrial, and agricultural territories (Ebenezer Howard, 2010). His quest for creating a pleasant

N. Toleva-Nowak (✉)
University of Architecture, Civil Engineering and Geodesy, Sofia, Bulgaria

semi-urban/semi-rural environment, providing alternative employment of the farming type in exchange for overcrowded unhealthy cities, emerged as a counter-reaction of the many severe problems caused by active migration and urbanization as well as the progressive deterioration of the urban microclimate (Ebenezer Howard, 2013). Ever since, his works have become a source of inspiration for a number of similar concepts and movements in architecture and urban planning, such as New Urbanism, Intelligent Urbanism, Urban Village, the concepts of Manuel de Landa, Japanese Metabolism, etc.

Nowadays, an increasing interest in urban agriculture is observed – both in developing and in developed countries. Among the many publications and analyses on the topic, a proof for this growing trend was the theme of the 2015 World EXPO in Milan, "Feeding the planet. Energy for Life" (Assolombarda, 2015), which focused on rediscovering the old traditions, interpreting the future of agriculture while keeping in mind the cultural background specifics, and food production in the context of a highly urbanized, technological and digitalized world. This theme was also reflected in one of the leading topics of Matera – European Capital of Culture in 2019 – the "Gardentopia project" which aimed to renovate abandoned urban spaces with the methods of urban agriculture and the cooperation of the local community (Matera Basilicata Foundation, 2019).

In the Balkan region and historical context, urban agriculture practices have even deeper roots. Its origins can be traced back to the first third of the nineteenth century with the beginning of the social transformation of the Bulgarian Revival. In order to understand better the local context of these activities and their spatial aspect, Revival yards in the towns of Kotel, Karlovo, Koprivshtitsa and Plovdiv, and their transition from the Early to the Mature Revival, were examined. The selected yards were located in an urban environment and there is evidence that they have preserved their authentic vegetation, or that their traditional maintenance has been continued over the centuries, which not only guarantees that the current vegetation is as close as possible to the original one, but also indicates strong historically sustainable practice. The historical study of the spatial development of urban agriculture was based on "The First Gardens of Plovdiv" by Boris Puhalev (1984), "Yard Gardens in Koprivshtitsa" by Rashko Robev (1981), "Karlovo Yards during the Renaissance" by Delcho Sugarev (1956) and "Renaissance Yards in Kotlensko" by Hristo Dimov (1976). The ratio of built-up area in the property, soil cover and pavements were examined. The preferred vegetation, its bio-diversity and the predominance of decorative and utilitarian plants were traced, taking into account the overall organization of the domestic yard in terms of function, sunshine, altitude and orientation to the cardinal directions and access from the street.

In the Renaissance, the development of crafts and trade led to a rapid economic and cultural upsurge, to a change in the social aspect and to intensifying processes of urbanization and migration to the cities. This transformed the whole organization of daily life, the understanding of comfort and placed new demands on the functional organization of the home and its adjacent spaces. Very specific feature of the Bulgarian Revival city is that each property is like a capsule, isolated from the public and street space by a high wall, thus the urban ensemble becomes a dissipative system in which the whole cell is oriented inwards and is organized by the

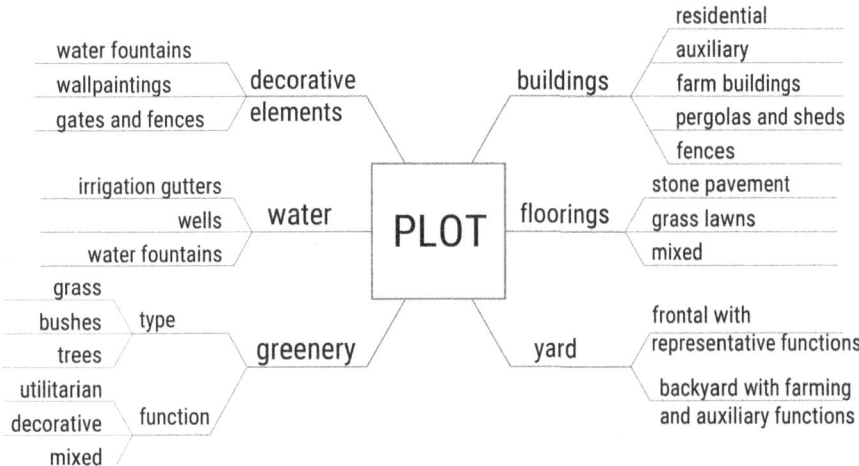

Fig. 7.1 Elements of the domestic yards

interactions of the functional and spatial elements, gathered in six main groups: buildings, floorings, yard, greenery, water and decorative elements (Fig. 7.1). The main accents of the Revival urban yard – representative lawns and flower beds, vegetable garden, vine arbour or giant walnut tree, farmyard with secondary buildings and sheds, are subject to specific functional zoning, depending on the location of the village, altitude, orientation towards cardinal directions, and terrain specifics, the traditional livelihood, natural resources, climate and soils.

The evolution of spatial organization follows the changes in the urban way of life as the elements do not significantly change their role or interrelations.

Architecturally speaking, during the Early Renaissance (late eighteenth century to the 1830s), outbuildings were traditionally associated with agriculture and crafts. The yard has a predominantly economic function, but the start of the formation of a representative area can be noticed. During the Mature Revival (1830s–1860s) the additional buildings were subordinated to handicrafts, trade and additional household functions (cooking, laundry, bathroom, storage, etc.). The front/representative and back farmyard are clearly defined. During the Late Revival (from the 1860s to the Liberation), the yard space evolved, the number of household outbuildings increased. The division of the front (representative) and back (household) yard continues, but the emphasis moved to the decorative and representative nature of the front yard, under Western European influences, refracted through the prism of the Ottoman Baroque. The main feature of Bulgarian yard gardening is its widespread practice, which favours a change in the overall attitude towards the urban environment and gives an initial impetus to the development of the hitherto non-existent urban green system.

The Revival yard is a fluid space, a container for the various daily activities. The yard composition is subordinated to the production processes, as the free space is limited by the house and the outbuildings. Even in the very tiny yards there are areas for flowers and utilitarian vegetation, as the composition of the main and additional buildings is subordinated to the search for optimal sunlight.

In addition to the cobblestones, the house, the production buildings, the flower beds, the vegetable garden, the orchard and the vine arbour traditional for the Revival yard, are visually and functionally separated. An important factor is the abundance of running water and the associated yard elements, accented with lavish ornaments – wells, fountains and gutters. With the transition from the Early to the Mature Period, the Revival yard unfolded all aspects of urban agriculture and practically turned into its prototype. Subsequently, in the period of the Late Revival, with an increasing number of auxiliary buildings, the space allocated for a garden also expanded proportionally, as its representative function matured. An interesting correlation in the larger yards in Karlovo and Koprivshtitsa was observed: even though the trend for the utilitarian function of plant species prevailed, larger areas were designated for purely decorative species (Toleva-Nowak, 2019). Even though the organization of the Revival yard varies from settlement to settlement due to the different geographic location, climate, altitude, the occupation of the owner and the traditional livelihoods in the area, the size and the overall shape of the plot and the orientation towards the cardinal directions, there are common functional interrelations between the elements of the Mature Revival yard (Fig. 7.2).

Urban agriculture has deep and sustainable roots in the Balkan context. Following the natural evolution of the vernacular architecture and the connection with the rural and agricultural lifestyle, it has not lost its relevance in the urban context. On the contrary, it saturates the urban environment and lays the foundation for the first urban landscaping and agriculture practices. Thus we may conclude that urban agriculture is not a new phenomenon imported into Bulgaria, but rather a traditional activity associated with improving the quality of life, enhancing the living environment and becoming a symbol of high social status, turning the house and the yard into a rhizome.

The aim of the research was to investigate the spatial aspects and the possible effects of urban agriculture practices on the improvement of the Quality of Life of the local communities in the City of Sofia. In order to reflect the complex nature of the urban agriculture practices in the contemporary capital city, a special methodology encompassing a multilevel approach was designed.

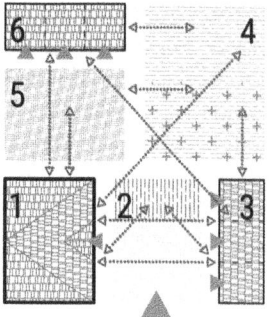

Legend:
1 - house
2 - front yard with representative function, often with vine arbour
3 - outbuildings for household needs (kitchen, bathroom, storage)
4 - vegetable garden/orchard
5 - back yard for occupational or secondary functions
6 - outbuildings for occupational or secondary functions
◁┄┄┄▷ - functional interrelations

Fig. 7.2 Functional scheme of the organization and interrelations in the Mature Revival yard

7.2 Methodology: Desk Research

During the first phase of the project from 2017 to 2018, desk research was conducted covering various historical, theoretical and regulatory aspects. It included selecting, reviewing and analysing specialized literature, gathering historical data, archive maps, architectural and urban plans and movements, contemporary urban agriculture practices and projects, and drawing conclusions about their multifunctionality and relevance. The desk research also included a historical overview of the proto-urban agriculture practices in Bulgaria during the Revival. It served as a foundation for the analysis of the current regulations and development strategies in order to identify the main obstacles to practising urban agriculture in contemporary Sofia. Last, but not least, the online presence of urban agriculture practices was traced via social media pages, publications, etc. The second phase of the spatial study (2018–2020) related to extensive and thorough field work, exploring the urban and architectural specifics of three very distinct types of urban environment in Sofia, which will be discussed in detail below.

The three districts presented in Fig. 7.3 were studied in regard to their location within the urban organism of the capital city, their morphology and connectivity, the predominant types of buildings, population density, building parameter regulations and zones. The field survey covered an overall area of 7327 km^2 and more than 202,500 inhabitants:

Using the historical overview as a background, several types of residents' interventions in the environment were selected as signs of conscious and subconscious urban agriculture practices: maintained grass areas and greenery gardens, flower beds, vegetable gardens, greenhouses, vines and vine arbours, fruit trees or orchards, beehives and DIY shared spaces. The data were mapped in their GPS coordinates during the field trips.

While in some places the field-trips were sufficient to document the interventions, in others they were less productive. On the one hand, it was practically impossible to enter each and every property and describe the family practices, while on the other it raised some ethical considerations. Thus, in all regions, along with the field surveys a satellite-image-based analysis and mapping were conducted, using open data from Google satellites, Landsat-8 and Sentinel-2. This method was preferred as it was based on overall land-use pattern analysis and, as it lacked detail, offered at least some level of anonymity.

It is important to note that the data do not claim complete accuracy, but rather represent an approximate number and types of interventions. Also, as urban agriculture practices are dynamic and change on a seasonal and annual basis, the mapping is more of a snapshot of the current state, which may change in the future.

The data were presented as multilayer interactive maps for open access via free Google-based software "My Maps". The method used was entirely descriptive.

In the final phase of analysis, the results showed how urban agriculture practices differ depending on the specifics of the district. Concerning the predefined Quality of Life indicators from an urban perspective (growing food and producing non-edible resources, providing healthy environment and promoting active lifestyle, improving

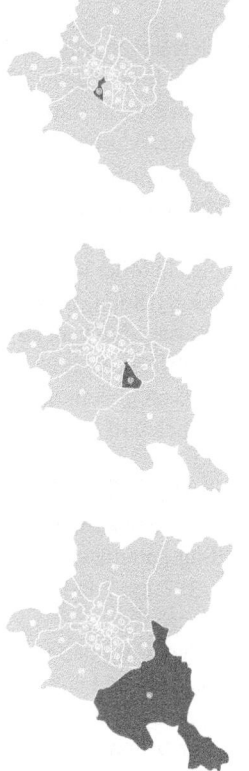

Triaditsa – a classical urban district, representing an almost perfect slice of the city extending from the centre up to the city ring-road, encompassing a variety of residential and public zones.

Mladost – a peri-urban district consisting of 1970s huge pre-fab residential buildings and the former village of Gorublyane with its single-family houses and green yards.

Pancharevo – a suburban district including 10 rural settlements and their adjacent territories (plains, agricultural lands, mountain ranges, forests, rivers, lakes, dams, pastures, etc.).

Fig. 7.3 Location and description of the three districts studied

social cohesion through stimulating spaces and opportunities for recreation), the produced statistics, graphics and maps supported the hypothesis that the centuries-old bottom-up traditions enhanced with contemporary socio-cultural practices are valid approaches for improving quality of life on the individual, communal and city level.

7.3   Study Results

The results from the field research and the satellite image analysis for Triaditsa, Mladost and Pancharevo regions show that on the individual property level, the traditional spatial organization of the Revival yard is preserved up to this day. As expected, the proportion of utilitarian vs decorative areas may vary according to the type of urban zone and the specifics of the settlement, yet the spatial arrangement is still the same – it depends not only on the access from the urban street network but also on the altitude and the orientation towards cardinal directions.

7.3.1   Triaditsa

Triaditsa is a central district, covering 9.8 km², with approximately 75,000 inhabitants, making an average density of 7563 people/km². It is of particular interest as it is an almost perfect slice of the urban organism – a cone stretching from the heart of the city up to the ring-road. It consists of a wide range of different urban zone types with different regulations and restrictions (Fig. 7.4), enriched with a wide variety of residential areas with different intensity and height (Fig. 7.5). The focus of attention in the research were the central mixed-use zones and the residential ones, which constitute approximately 50% of Triaditsa territory.

The General Development Plan for Sofia Municipality is the foundation for urban development not only in the short, but in the mid- and long-term. It gives very precise definitions of the different types of zones (public, residential, recreational,

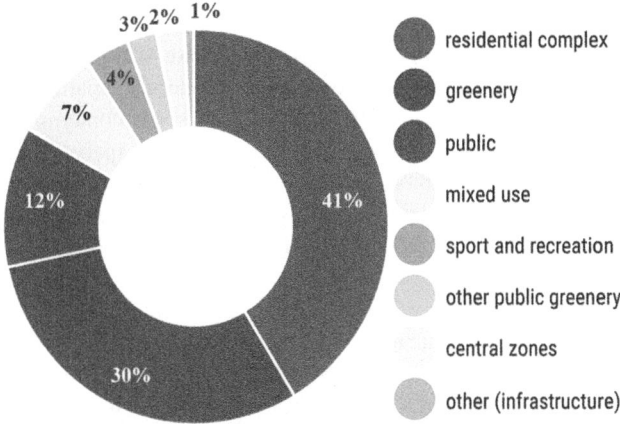

Fig. 7.4 Urban zones types in accordance with the General Development Plan (GDP)

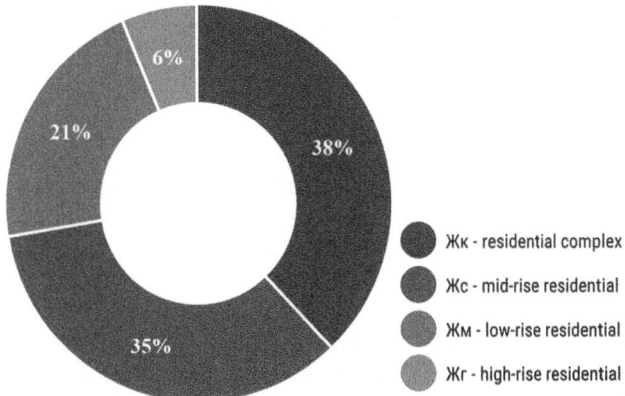

Fig. 7.5 Proportion of the different residential zone types according to the GDP

mixed-use, industrial, agricultural and technical infrastructure zones, etc.) with their specific parameters for the maximal height, maximum permitted built-up area, minimal required landscaping areas, protected areas, distances between the edifices and infrastructural servitudes. There are approximately 70 different types of zones, a dozen of which are residential. However, in Triaditsa, for the purposes of the study, we may colligate the wide variety of different types of residential zones into three main groups:

- **Central mixed-use parts** with mostly medium-density terraced houses, with a neat facade front (buildings with an average height of about 6 floors, built around the 1960s, after the Second World War, housing public service functions in the ground floors and apartments in the upper ones). They enclose inner-quarter spaces, inaccessible for the public and the ordinary pedestrian, filled with small green patches, trees and sometimes parking space. Along with these structures, there are a few family houses with beautiful frontal yards, built mostly before the Second World War (some of them even at the beginning of the twentieth century, yet they are rather an exception).
- **The residential complex and areas for high-rise residential buildings** are characterized by the construction of pre-fab modernist apartments buildings, constructed in the 1970s and 1980s, as well as some contemporary edifices. With heights from 6 to 16 floors and vast in-between spaces, filled with rich greenery and grass lawns, plenty of high trees, parking lots and children playgrounds, these territories constitute up to 44% of Triaditsa territory. For them, the General Development Plan provides some restrictions (see Table 7.1).
- **The suburban area** comprises of mid-rise residential buildings and predominantly low-rise single-family houses with neatly arranged domestic yards and traditional functional organization. The area includes several types of low-rise residential zones with varying density, intensity coefficient, minimum greenery percentage and maximum allowed height (Table 7.1). Some low-rise residential zones (Жм2 and Жм3) with very strict construction restrictions are situated towards the outskirts, adjacent to the ring-road.

For the whole Triaditsa District overall, 707 interventions were mapped: 153 vegetable gardens, 19 greenhouses, 84 vine arbours, 5 orchards, 430 green lawn patches, 22 flower gardens, 6 apiaries and 9 green roofs (Fig 7.7). Nearly 69% of the interventions were located in purely residential areas and 13% in mixed-use areas. Particularly interesting is that approximately 14% of the interventions are located in public green areas (Fig. 7.6). As expected, most of the traces of urban agriculture practices were well-maintained lawns – 59%, followed by vegetable gardens (21%), and vines and vine arbours (11%) (Fig. 7.7).

Noteworthy is the proportion of the different types of interventions in the different zones (Figs. 7.8 , 7.9 and 7.10).

The spatial distribution of the different types of intervention shows that the intense mixed-use residential/administrative territories are less saturated with urban agriculture practices, and the few interventions are mostly decorative with little need for maintenance and care – small green grass patches, some few flower

Table 7.1 Construction parameters in residential zones according to the General Development Plan in Sofia

Zone type	Character	Density of construction	Intensity coefficient	Minimum green area	Maximum height
Жг	High-rise residential				
Жк	Residential complex	40%	3	40%	26 m (residential) 31m (public)
Жс	Mid-rise residential	50%	2.3	35%	15m (residential) 20m (public)
Жм	Low-rise residential	40%	1.3	40%	10m (residential)
Жм1	Low-rise residential in natural environment	30%	0.9	60%	10m (residential)
Жм2	Low-rise construction and additional specific requirements	30%	1	60%	10m (residential)
Жм3	Residential area with low-rise buildings with restrictive parameters	15%	0.3	80%	7.5 (residential)

Fig. 7.6 Proportion of interventions per zone type

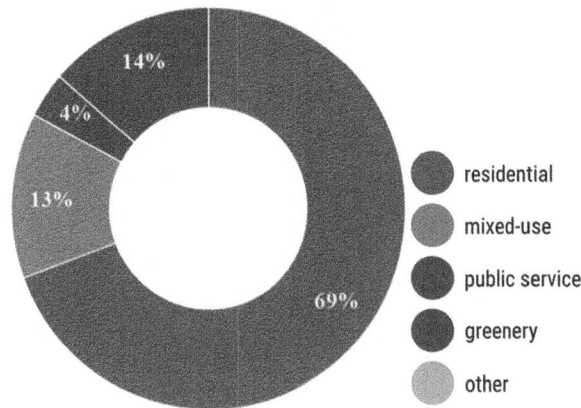

beds and very few vines (Fig. 7.11b). As expected near the centre of the city and in the residential complex areas, where the free space is public and shared between lots of inhabitants, the interventions are mostly decorative, while towards the outskirts, in the peri-urban territories filled with family houses with their own domestic yards, the utilitarian aspect prevails – vegetable gardens, vine arbours, even apiaries (Fig. 7.11c–f). We may claim, that the shift from decorative to utilitarian changes with the urban scale and the property ownership, conscious utilitarian agriculture is not practised within vast public open spaces (in the residential complexes there are plenty of green areas), but more within the privately owned yards, with lower building height, providing enough sunshine for the plants. This data corresponds with the findings from the socio-economic survey, which demonstrated that the predominant UA practice in Sofia is of private and family character and public

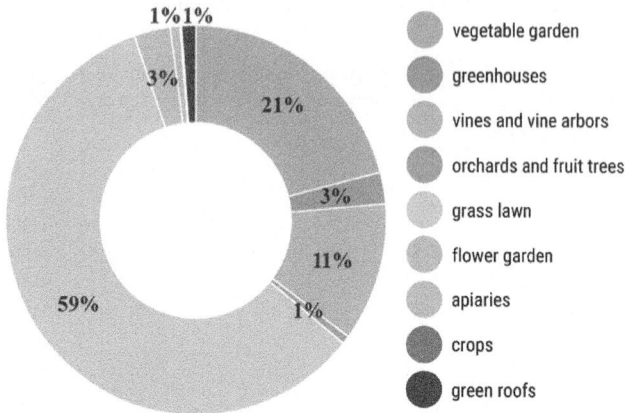

Fig. 7.7 Types of interventions in Triaditsa region

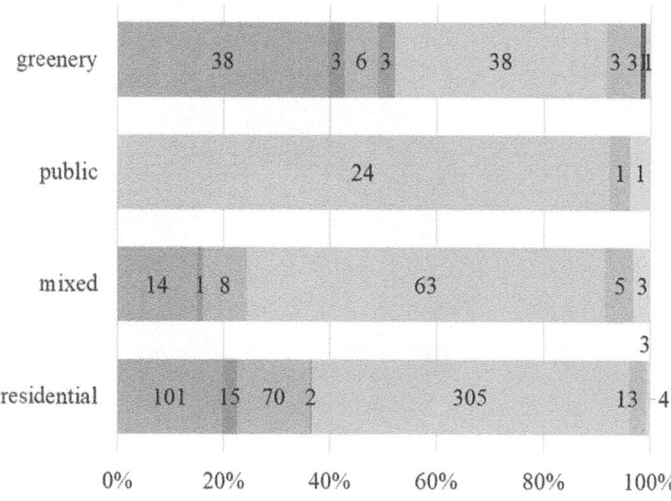

Fig. 7.8 Types of interventions in residential/mixed/public and green areas in Triaditsa

space is rarely utilized for collective communal activities and goals. Other factors are also taken into account – easy access to irrigation water, the types of soil, the air pollution, etc. Last but not least, a quite worrying trend was spotted – the closure of the green wedge (Fig. 7.11a), which threatens the whole functionality of the urban green system.

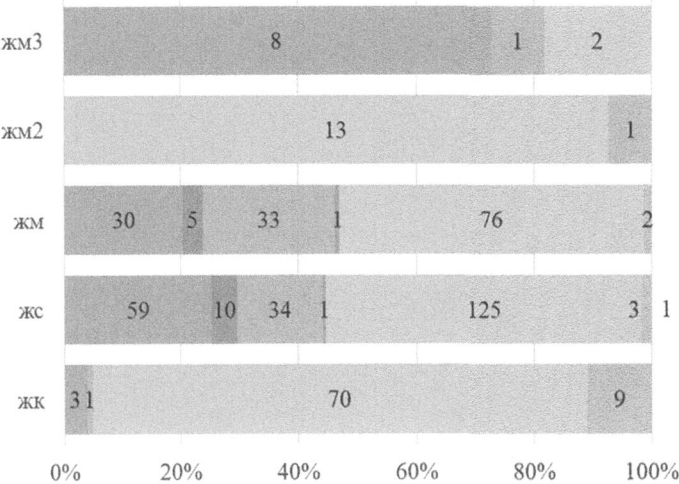

Fig. 7.9 Types of interventions in residential areas in Triaditsa

Fig. 7.10 QR code and link
for Triaditsa map
https://cutt.ly/LcSYl9o

7.3.2 Mladost

The Mladost District covers several sub-regions: the Mladost Residential Complexes 1, 1A, 2, 3, 4, the areas of the American College in Sofia, Polygona and the former village of Gorublyane. The region extends over approximately 16.78km², with a population of 117,112 people, thus reaching an average density of 6979 people/km². All the sub-regions, except Polygona which consists of large industrial or office buildings, have a predominantly residential character. There is a variety of urban zone types within the district – 24% residential, 24% mixed-use (residential and public service), 26% are greenery, and the rest are for public, sport, recreation, technical infrastructure and others purposes (Fig. 7.12a). The types of residential territory vary quite a lot: 55% are high-rise residential complexes (Жк), 42% – low-rise residential areas (Жм) for family houses, 3% are mid-rise residential buildings (Жс) and the other less than 1% are the high-rise residential territories

Fig. 7.11 (**a**) zone types, (**b**) all interventions, (**c**) vegetable gardens and greenhouses, (**d**) vines, arbours, fruit trees and orchards, (**e**) grass lawns, (**f**) flower gardens and flower beds (**g**) apiaries

(Жг) (Fig. 7.12b). The same construction restrictions as in Triaditsa District apply (see Table 7.1).

Mladost District is unique with its two very contrasting types of residential usage. The 1970s-1990s high-rise residential complexes with prefab apartment buildings and spacious inter-block areas, housing parking lots, children's playgrounds, and neglected and often polluted muddy grass lawns. A major problem is the lack of enough parking space and designated ground or underground parking lots, forcing the residents to use the green areas and the sidewalks for this purpose, creating an overcrowded, polluted and unsafe environment. Furthermore, this not only worsens the air quality, but disfigures the visual qualities of the environment and prevents the people from enjoying the greenery and the nature within the city – one of the keystones of modernist urbanism. On the other hand, there are multiple attempts at bettering the immediate environment made by the inhabitants of the apartment buildings – maintaining the green patches in front of the building entrance, constructing gazebos, vine arbours, DIY children-friendly facilities and play-grounds. In contrast with this image are the low-rise residential areas of the

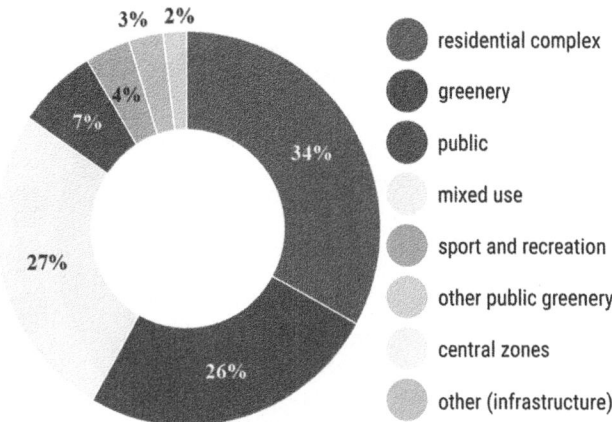

Fig. 7.12a The variety of urban zones in Mladost District

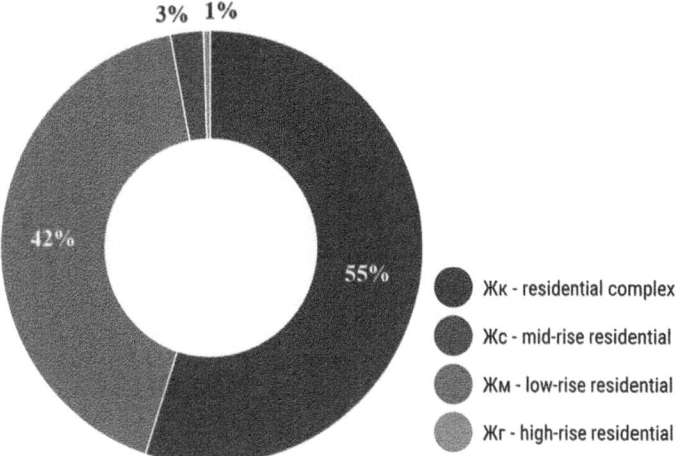

Fig. 7.12b Types of residential areas in Mladost District

neighbourhood around one of the oldest foreign college institutions – The American College, and Gorublyane – a former separate settlement of rural type, where the predominant buildings are single or multiple family houses with their own domestic yards, narrower streets and lavish greenery.

Overall, 2869 interventions were identified in Mladost District: 459 vegetable gardens, 151 greenhouses, 759 vine arbours, 1045 green grass patches, 4 apiaries and a number of single fruit trees and some orchards (Fig. 7.14a). Due to the specifics of the region (as the so-called urban bedroom), not surprisingly 88% of the interventions occur in the residential zones and only 10% in the mixed-use areas (Fig. 7.13). As in Triaditsa District, we observe some interventions in the public

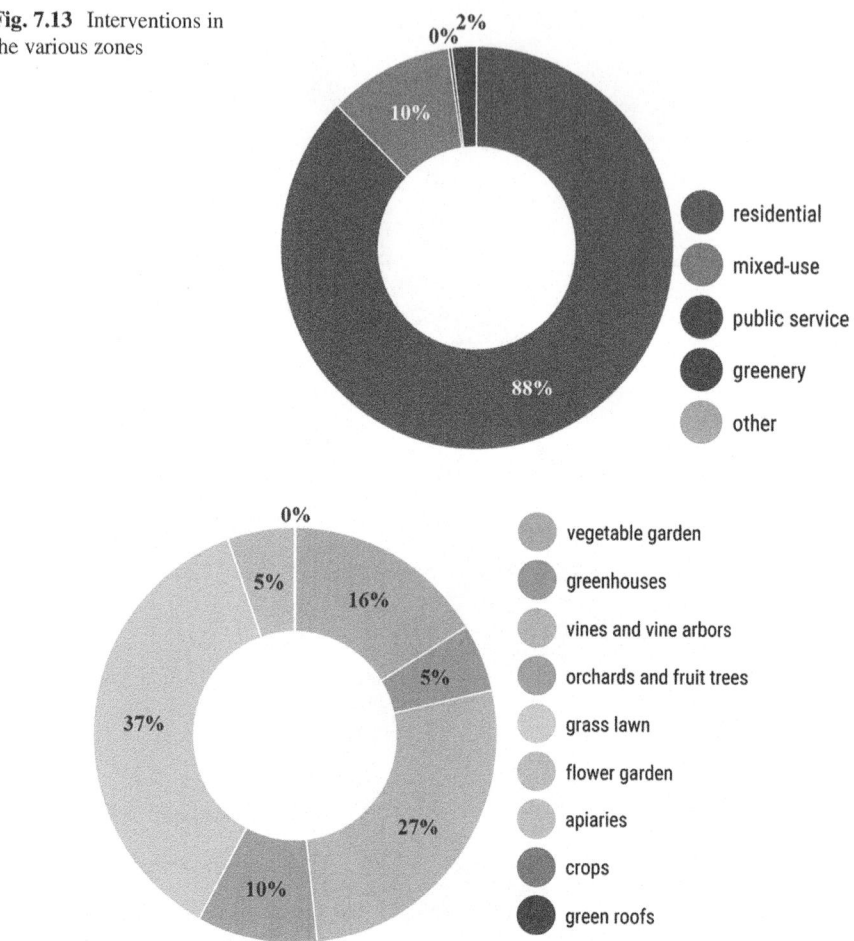

Fig. 7.13 Interventions in the various zones

residential
mixed-use
public service
greenery
other

vegetable garden
greenhouses
vines and vine arbors
orchards and fruit trees
grass lawn
flower garden
apiaries
crops
green roofs

Fig. 7.14a Types of interventions in Mladost District

green areas. Impressive, yet not unexpected, was the change in the proportion of the various urban agriculture interventions in the three sub-regions – Mladost, American College and Gorublyane (Figs. 7.14a and 7.14b). While in the residential complexes the decorative aspect prevails, in American College area the trend starts to shift with the emergence of some vegetable gardens and greenhouses, and finally in Gorublyane the utilitarian aspect prevails – more than 1/3 of the interventions are vegetable houses and greenhouses, 1/3 are vines and vine arbours and the rest are grass lawns and flower beds (Fig. 7.14c).

This transition is very visible when we look at the spatial distribution within the district (Fig. 7.15). Particularly interesting is the intensity of the interventions in Gorublyane – 56% of all interventions in an area with just under 6% of the population of the district, making an average urban agriculture activity of

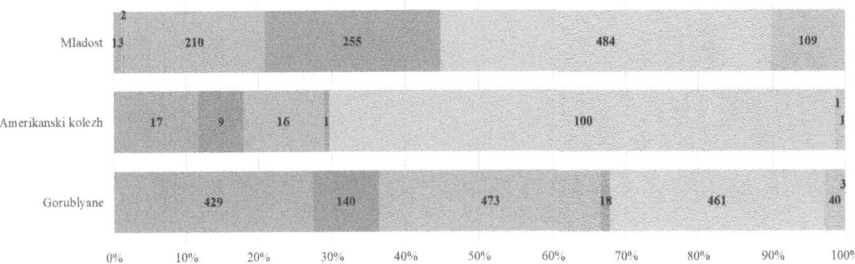

Fig. 7.14b Comparison between the proportion of the various interventions in the three sub-regions of Mladost District: Mladost residential complex (top), American College area (middle), Gorublyane former village (bottom)

Fig. 7.14c QR code and link for Mladost and Gorublyane map
https://cutt.ly/9cSUqIf
https://cutt.ly/HcSUT4t

22 interventions per 100 people. Luckily, here the danger of over-construction in the green wedges is still under control.

We have to acknowledge some of the issues here. As the urban environment is very different in terms of scale, intensity, shared-use and spatial characteristics, the value of some of the urban agriculture activities is not the same. While in almost each house yard there is a small flower bed, and in a few of them there are bigger flower gardens, in the adjacent areas near the huge apartment buildings even the small flower beds signify active participation and have high value, due to the difficulties in their maintenance (often the residents have to go to their apartment for tools or water for irrigation). Thus, in the low-rise residential areas, only the bigger flower gardens were mapped, while in high-rise and residential complex areas even the small flower beds were traced.

The same argument is applied in tracing the fruit trees and the orchards. Astonishingly, not only the planting of new fruit trees (cherries, sour-cherries, peaches, hazelnuts, vines, etc), but also the care for self-grown fruit trees (plums, mulberries, walnuts) – limewashing, protecting with fences, pruning and hoeing. Thus, like the flower gardens, the value of a single fruit tree in shared space in a residential

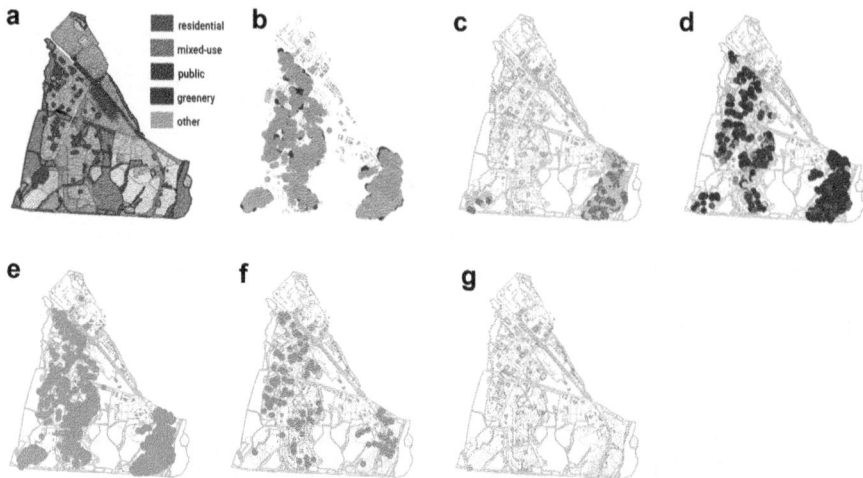

Fig. 7.15 (**a**) zones, (**b**) grass and flowers, (**c**) fruits and orchards, (**d**) vines and arbours, (**d**) vegetables and greenhouses, (**e**) apiaries

complex may correspond to an orchard in an individual plot, as in the domestic yard almost all trees have a utilitarian character (Fig. 7.15c). An impressive amount of the vines and vine arbours are located in the residential complex areas – more than 28% of all traced vine plants (Fig. 7.15d). A small side-study is presented later in the Analysis section, exploring the functional and the decorative nature of the vine-growing practices in a high-rise residential complex environment (Figs. 7.36 and 7.37).

7.3.3 Pancharevo Region

Pancharevo is a suburban district with an overall area of 332.62 km² and a population of 27,629 people, resulting in a relatively low density of 83 people/km². It consists of 10 rural settlements (Bistritsa, Pancharevo, Kokalyane, German, Zheleznitsa, Lozen, Kazichene, Krivina, Plana and Dolni Pasarel) and their adjacent agricultural lands, forests, pastures, dams, lakes, natural reserves, etc. The terrain is predominantly mountainous, smoothly transitioning into the Sofia Plain.

The survey was implemented in all the 10 settlements, mostly in their central territories, as some of the villa-type zones for seasonal habitation were hard to access, extremely scattered and with insufficient satellite data information.

7.3.4 Bistritsa

Bistritsa is located approximately 15 km south of Sofia, at the foot of Vitosha mountain. In 1998, by a decision of the Sofia Municipal Council, it was declared a resort area. It is one of the oldest villages in Sofia Region. Currently it has a population of approximately 5000 people and its lands extend over 52 km^2. Three rivers cross the settlement, defining its mountainous character and bountiful greenery. The main village is surrounded by nine luxury low-rise villa zones, intended for seasonal exploitation in a natural environment, as well as some newly erected luxury gated residential complexes.

On the territory of the settlement and part of its adjacent villa zones 1851 interventions were identified, distributed by type as follows: 422 vegetable gardens, 174 greenhouses, 328 vine arbours, 74 orchards, 769 green spots, 73 flower gardens, 9 apiaries, 2 agricultural plots within the village (the agricultural territories tangential to the settlement were not taken into account) (Fig. 7.16b), which results in 37 interventions per 100 people.

Most of the urban agriculture interventions (97%) were located in the residential zones (Figs. 7.16a and 7.16b) and had a predominantly decorative aspect – 42% were maintained grass lawns and 4% large flower gardens (the tiny flower beds were not taken into account as they can be observed in almost every house in the village region and did not require special care). On the utilitarian side – 23% of the interventions were vegetable gardens and 9% greenhouses. The vines and vine arbours constitute 18% of all interventions (Fig. 7.16c). There are no significant discrepancies in the spatial distribution throughout the various residential region zones (Fig. 7.16d) (The zone restrictions are the same as listed in Table 7.1) (Fig. 7.16e).

Of interest is the concentration of the utilitarian interventions (vegetable gardens and greenhouses) in the central urban area: this in contrast with the higher density of maintained lawns is in the north-eastern part of the settlement where the luxury gated

Fig. 7.16a Interventions per zone types

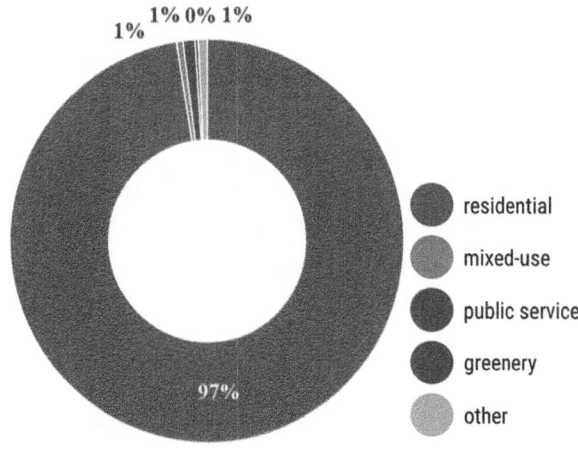

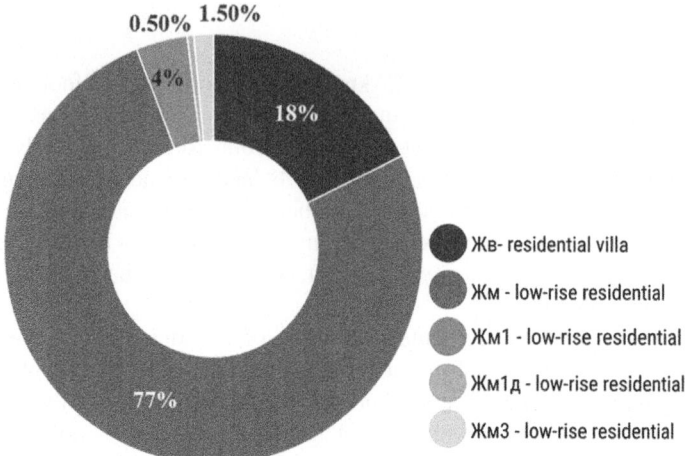

Fig. 7.16b Types of residential territories in Bistritsa

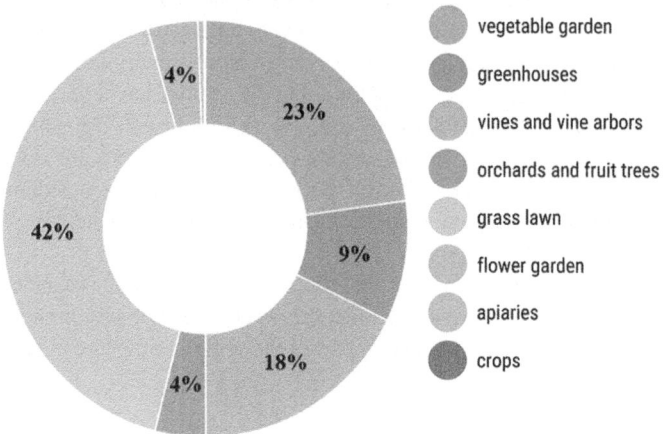

Fig. 7.16c Interventions in Bistritsa by type

complexes are located. The apiaries are towards the outskirts. In the General Development Plan a major extension of urban residential zones over agricultural and forest lands is planned (Fig. 7.17).

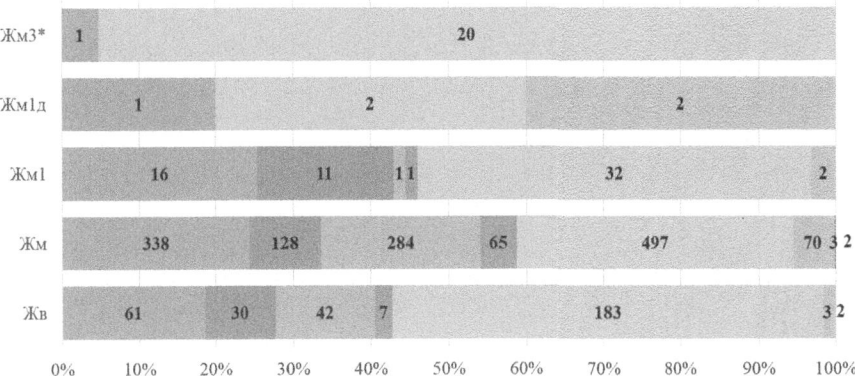

Fig. 7.16d Interventions in the different residential zones

Fig. 7.16e QR code and
link for Bistritsa map
https://cutt.ly/jcSUNhC

7.3.5 German

German is located approximately 12 km south-east of Sofia, on the east side of the
Iskar River – a major factor for the morphological development and spatial evolution
of the settlement. Its lands cover approximately 17 km^2 and it has nearly 2500
inhabitants. There are three neighbourhoods – Gorni (Upper), Sredni (Middle) and
Dolni (Lower) German. It is widely believed that the etymology of the name of the
village comes from the ancient Thracian word "germa", which means "hot spring",
as there are multiple hot springs south of the village. To the north, the adjacent lands
are characterized by fertile soils and the terrain smoothly spills into the Sofia Plain,
where vast agricultural lands and orchards extend. The terrain of Sredni German is
rather hilly, unique with its picturesque valleys and fields, while the southern part is
entirely mountainous and covered with dense oak, beech, hornbeam and hazel
forests, complemented by numerous meadows and pastures.

During the research only the main settlement was explored, as the extensive villa
zones located on the eastern slopes were extremely dispersed, with huge individual
plots, difficult access and used mostly for seasonal and weekend habitation. A total

Fig. 7.17 (**a**) zone types, (**b**) vegetable gardens and greenhouses, (**c**) vines, vine arbours and orchards, (**d**) maintained grass lawns and flower beds, (**e**) apiaries

Fig 7.18a Interventions per zone types

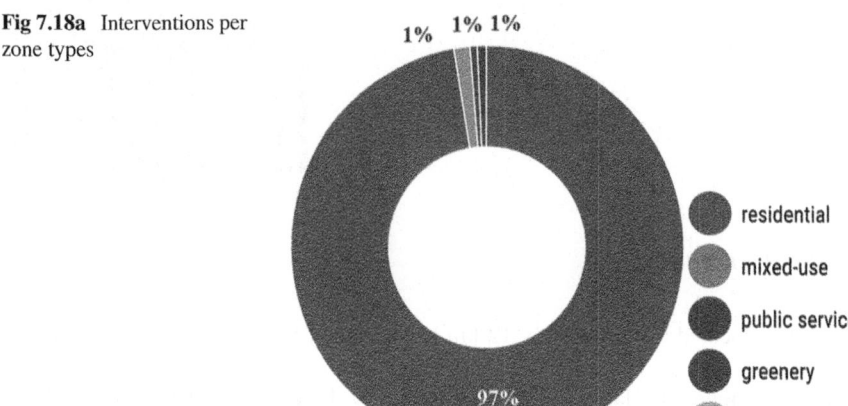

of 674 interventions were registered: 159 vegetable gardens, 53 greenhouses, 230 vine arbours, 31 orchards, 189 green plots, 9 flower gardens and 3 agricultural plots within the urban organism (Fig. 7.18b), making an average intensity of agricultural activities of 27 interventions per 100 capita. Again, almost all of the interventions occur in the residential areas (Figs. 7.18a, 7.18b and 7.18c).

After the end of the research, as part of some monitoring over the territories, a new intervention was spotted in the villa areas, most of which are maintained grass lands and vines with vine arbours. However, within the settlement the distribution of the various urban agriculture practices is quite homogenous and none of the

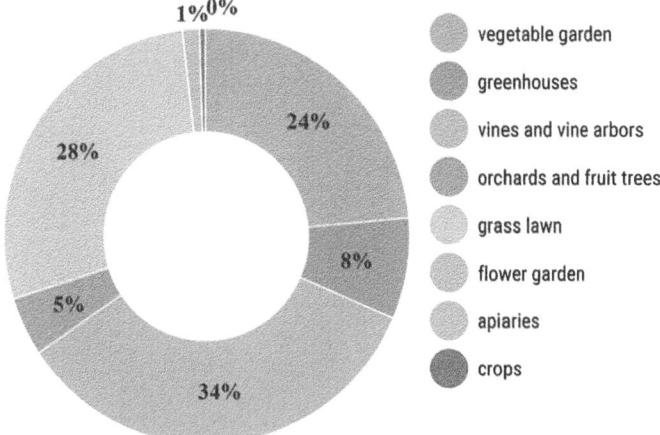

Fig 7.18b Interventions in German by type

Fig. 7.18c QR code and
link for German map
https://cutt.ly/mcSIphX

activities (vegetable gardening, maintaining grass lawns or taking care of the vines)
prevails significantly (Fig. 7.19).

7.3.6 Dolni Pasarel

Dolni Pasarel is a mountain village situated in a cosy valley between Plana and
Lozen mountains, approximately 13 km southeast of Sofia. Its lands extend over
52 km^2, and with it 1276 residents it has a rather low population density of 24.38
people/km^2. It is known for its rich biodiversity, numerous mountain springs,
beautiful protected areas and natural landmarks (Fig. 7.20a).

A total of 667 interventions were registered: 320 vegetable gardens, 155 greenhouses, 55 vine arbours, 7 orchards, 123 green plots, 4 flower garden, 3 apiaries
(Fig. 7.20b), 97% of which are in the residential areas (Fig. 7.20a). Surprisingly, just

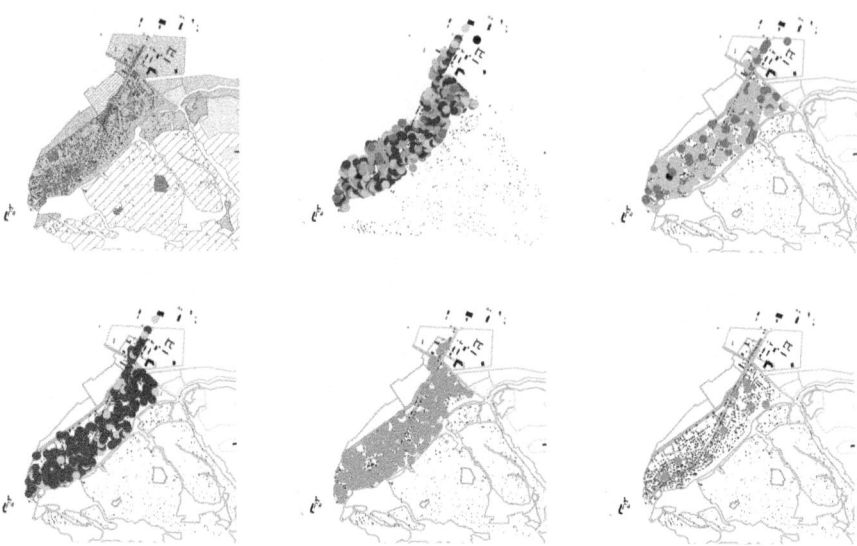

Fig. 7.19 (**a**) German zones (**b**) accumulation of interventions; (**c**) vegetable gardens and green-houses, (**d**) vines, vine arbours and orchards, (**e**) grass lawns, (**f**) flower gardens

Fig. 7.20a Interventions per zone type

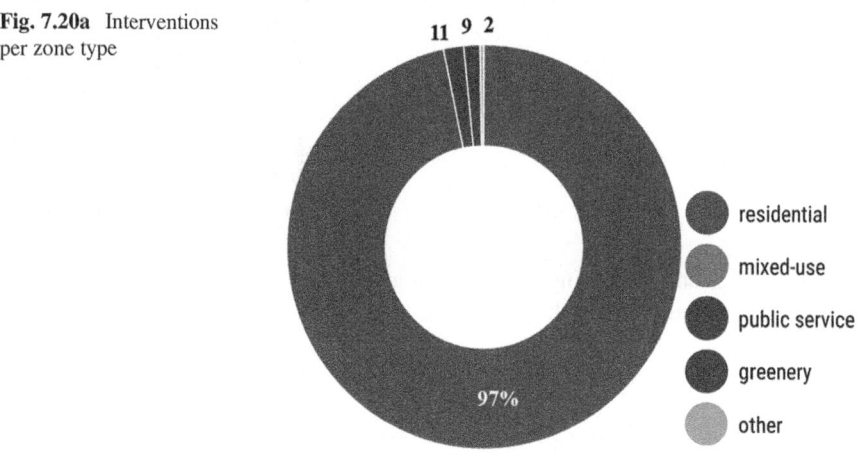

like Triaditsa and Mladost, some of the vegetable gardens and the greenhouses are located in the public green areas (Fig. 7.20c).

Being a mountainous village, it is not surprising that the vines constitute only 9% of the urban agriculture practices. What is striking, is the fact that almost half of all interventions are vegetable gardens, and a quarter of them are greenhouses, render-ing almost 75% of all interventions as completely utilitarian. Also, the proportion between the gardens and the greenhouses may be explained with the harsher mountainous climate and lower temperatures. However, the decorative aspect is

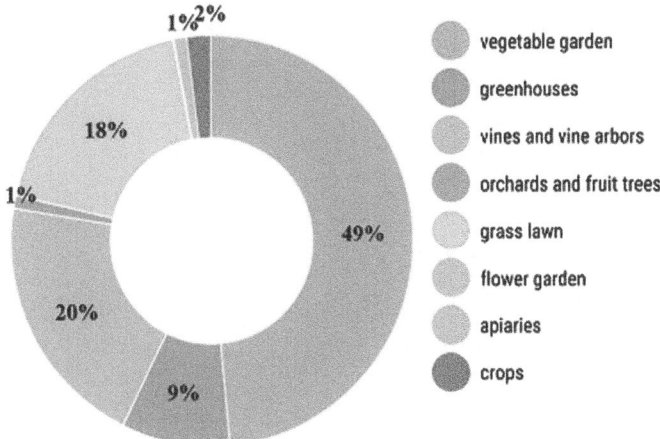

Fig. 7.20b Interventions in Dolni Pasarel by type

Fig. 7.20c QR code and
link for Dolni Pasarel map
https://cutt.ly/NcSIXKI

not completely missing – it is represented by evenly distributed neat grass lawns.
The intense utilitarian environment is topped with some apiaries on the north-eastern
side (Fig. 7.21).

7.3.7   Zheleznitsa

Zheleznitsa is a mountainous village located 23 km south of Sofia, on the eastern
slopes of Vitosha mountain. Its lands cover approximately 50 km^2 and it has 1586
inhabitants (31.53 people/km^2). The region has numerous mineral springs. The
settlement morphology is entirely subordinated to the direction of Vedena River,
which divides the urban structure. Extensive villa areas spread to the north and to
the east.

Fig. 7.21 (**a**) zones, (**b**) interventions, (**c**) vegetable gardens and greenhouses, (**d**) vines, vine arbours, orchards, (**e**) grass lawns and flower gardens, (**f**) apiaries

Fig. 7.22a Interventions per zone type

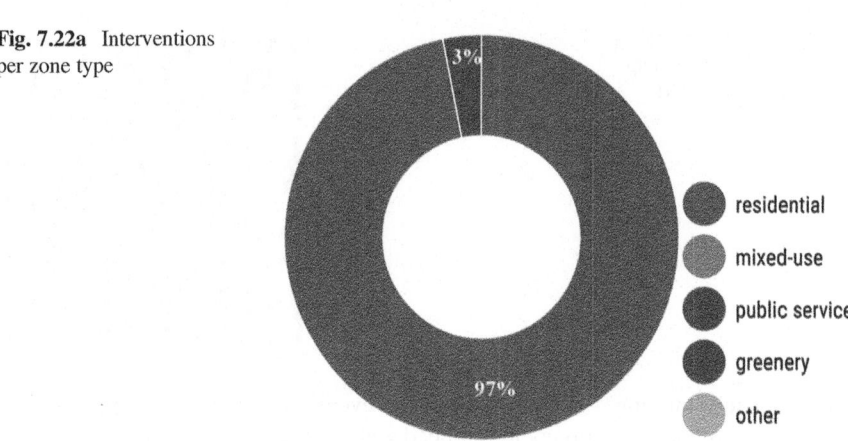

On the territory of the settlement and part of its adjacent zones 407 interventions have been registered: 122 vegetable gardens, 59 greenhouses, 84 vines and vine arbours, 26 orchards, 94 maintained grass lawns and 22 large flower gardens (Fig. 7.22b). Most of the interventions were located in residential zones (Figs. 7.22a and 7.22b), and once again the use of public green areas for private urban agriculture practices is observed (Figs. 7.22c and 7.22d).

Again, the future development plan maps out massive residential zones over existing agricultural areas (Fig. 7.23a). The various urban agriculture practices are homogenous, yet not very intense (approximately 26 interventions per 100 people). There is a slight disproportion in the distribution of the utilitarian practices through-out the settlement fabric – the vegetable gardens and the greenhouses are denser in the centre of the village, while the maintained grass lawns prevail towards the southern outskirts. (Fig. 7.23b–d).

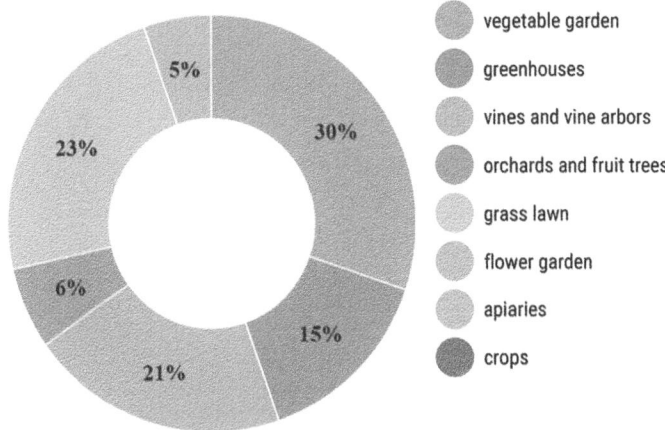

Fig. 7.22b Interventions in Zheleznitsa by type

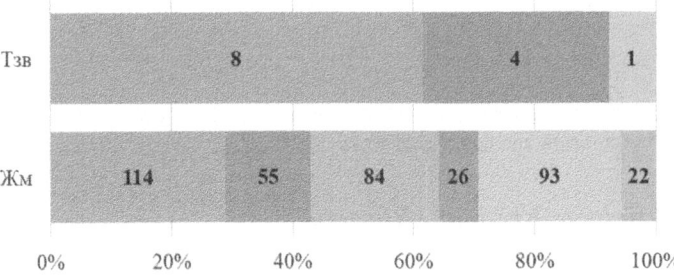

Fig. 7.22c Interventions: comparison between residential (Жм) and public (Тзв) zones

Fig. 7.22d QR code and link for Zheleznitsa map https://cutt.ly/ucSOaRE

Fig. 7.23 (**a**) zones, (**b**) vegetable gardens and greenhouses, (**c**) vines, vine arbours and orchards, (**d**) maintained grass lawns and flower gardens

7.3.8 Kazichene-Krivina – Rhizome

Unlike the above-mentioned settlement, Kazichene and Krivina are both located in the Sofia Plain. The two of them form a fully functional rhizome – their proximity and the similarities in the evolution of their structures resulted in increased interconnectedness, and thus they will be discussed as a dynamic whole.

Kazichene is located 12.5 km east of Sofia, north of the slopes of Lozen mountain. On its east side extend the fruitful agricultural lands of the Sofia Plain, while on its west side, after 2000, an industrial zone developed. In the recent years, an active population growth due to increased migration is observed, both due to the large number of newly created jobs and to the convenient proximity to Sofia City while enjoying the benefits and the intimacy of the nearby countryside. Currently, the population amounts to approximately 4800 permanent inhabitants, with a relatively high density of 290 people/km^2. The lands of the settlement cover approximately 17km^2. The active migration is leading to an increased demand for real estate properties, resulting, again, in plans for expansion to the east, over existing agricultural territories. A favourable factor for urban agriculture practices is the presence of mineral springs, used in the past for heating the greenhouses, as the thermal water quality was not good enough for direct consumption.

Fig. 7.24a Interventions
per zone type

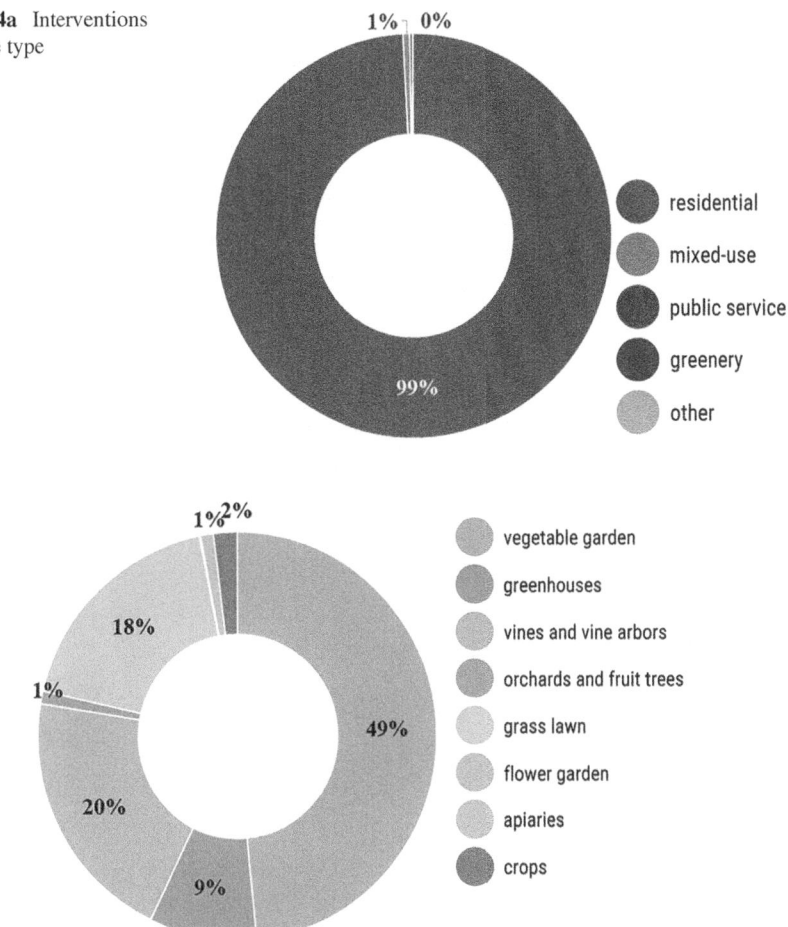

Fig. 7.24b Interventions in Kazichene by type

In Kazichene, a total of 1269 interventions were registered: 616 vegetable gardens, 111 greenhouses 258 vine arbours, 13 orchards, 233 green plots, 22 agricultural lands in the urban fabric, 13 apiaries and 1 vast flower garden (Fig. 7.24b). Nearly all of the interventions are in the residential zones (Fig. 7.24a).

The utilitarian aspect heavily prevails: almost half of the urban agriculture practices are related to producing vegetables (49% vegetable gardens and 9% greenhouses) (Fig. 7.24b). Unlike other villages, in Kazichene a good amount of the properties have high opaque fence, thus rendering the research of flower gardens almost impossible. The role of the vine arbour is particularly strong – creating a pleasant frontal shared space for the household to work outside or gather for a meal (Fig. 7.24c).

Fig. 7.24c QR code and
link for Kazichene map
https://cutt.ly/0cSOWkB

Fig. 7.25a Interventions
per zone type

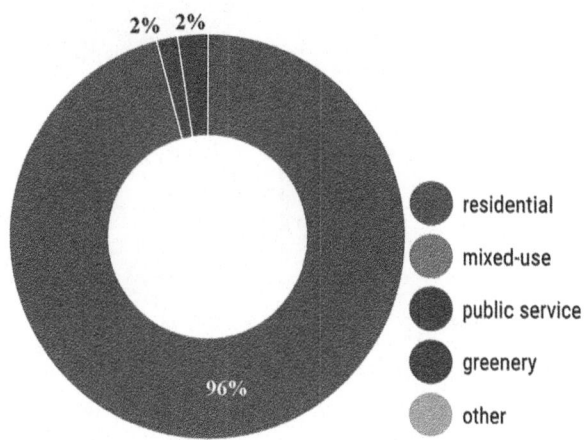

Krivina is located north of Kazichene, 11 km east of Sofia. It covers 11 km^2 and
has 1462 permanent residents (129 people/km^2). The adjacent territories are fertile,
rich in underground water, occupied mainly by agricultural land. To the east and to
the north, the village is bounded by lakes. To the west, its growth is restricted by the
ring-road. On the north side, there is also a big quarry for sand and gravel, which was
modernized in 2007. However, due to the increased migration and the lack of enough
real estate for the newcomers, the village is planned to expand only over agricultural
lands between the existing houses and the ring-road.

In Krivina overall 701 interventions were registered: 303 vegetable gardens,
86 greenhouses, 176 vine arbours, 7 orchards, 101 green plots, 4 apiaries, 24 agricultural plots within the urban boundaries (Fig. 7.25b), and almost all of them in the
residential zones (Figs. 7.25a, 7.25b and 7.25c).

Again, the utilitarian aspect prevails due to the favourable conditions and the
moderate-continental climate. What is specific is the decreased activity of
maintaining the grass lawns – only 14% of all mapped interventions, yet there is
an increased intensity of growing vines – 25% of all interventions (in comparison
with 20% in Kazichene) (Fig. 7.26).

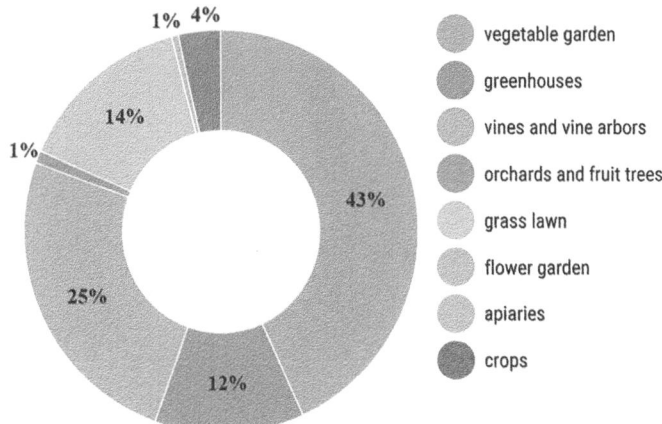

vegetable garden

greenhouses

vines and vine arbors

orchards and fruit trees

grass lawn

flower garden

apiaries

crops

Fig. 7.25b Interventions in Krivina by type

Fig. 7.25c QR code and
link for Krivina map
https://cutt.ly/ucSOXCz

7.3.9 Lozen

The village of Lozen consists of two parts – Gorni (Upper) Lozen and Dolni (Lower) Lozen, united in 1955. It is the second most populous village in Bulgaria (6080 residents, with an average density of 66.37 people/km^2 and an area of 91.6km^2), located 17.5 km southeast of Sofia centre, on the fringe of Lozen mountain. Its geographical location is a major factor: the prolonged hours of sunshine, which along with the fertile soils and the shallow groundwater, strongly favour urban agricultural practices.

On the territory of the settlement and part of its adjacent zones were registered 2168 interventions: 747 vegetable gardens, 203 greenhouses, 550 vine arbours, 130 orchards, 516 green plots, 4 apiaries, and 16 agricultural plots within the urban organism. Most of the interventions (98%) occur in the three types of low-rise residential zones (Fig. 7.27a) and the utilitarian aspect prevails (Fig. 7.27b). Interesting is the balance of the distribution of the interventions within the three residential zones – mostly utilitarian in the central zones, somewhat

Fig. 7.26 (**a**) zones (**b**) all interventions (**c**) vegetable gardens, greenhouses and crops, (**d**) vines, vine arbours and orchards, (**e**) grass lawns, (**f**) apiaries (Kazichene – bottom, Krivina – top)

Fig. 7.27a Interventions per zone type

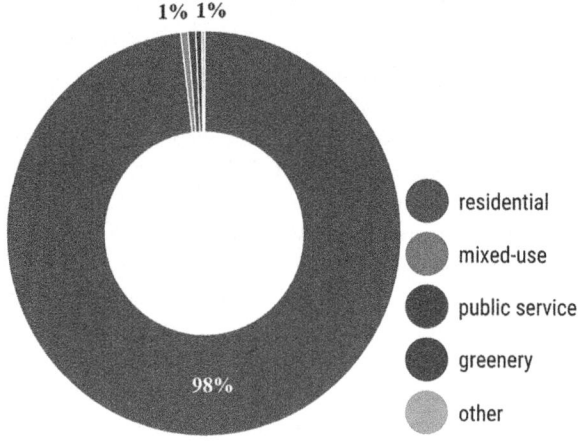

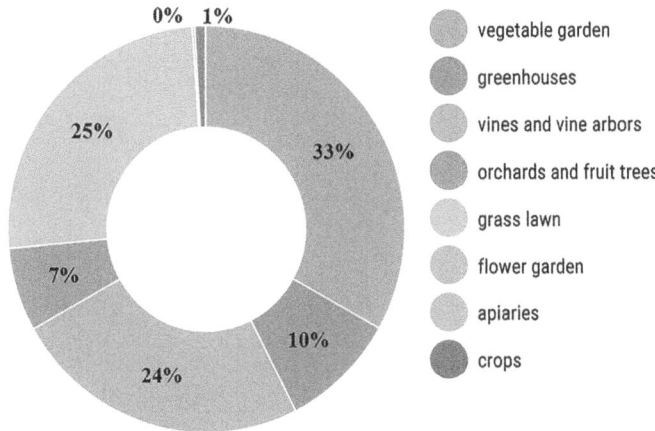

Fig. 7.27b Interventions in Lozen by type

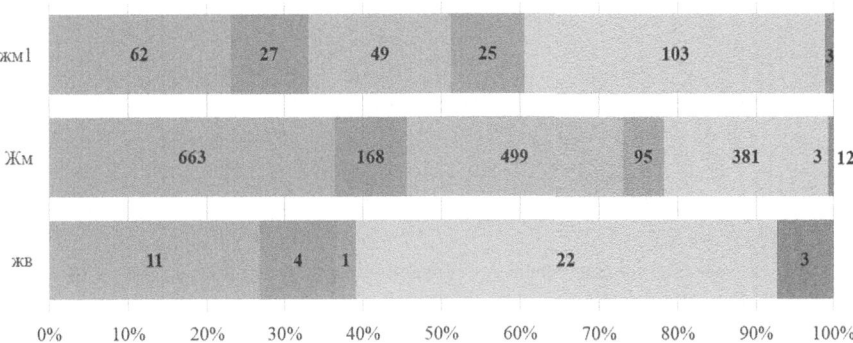

Fig. 7.27c Interventions comparison three different residential zones: low-rise residential with restrictions Жм1, low-rise residential Жм and residential villa Жв (see Table 7.1)

Fig. 7.27d QR code and
link for Lozen map
https://cutt.ly/OcSPsai

Fig. 7.28 (a) zones (b) all interventions (c) vegetable gardens, greenhouses and crops, (d) vines, vine arbours and orchards, (e) grass lawns, (f) apiaries

balanced in the transitional territories and mostly decorative in the villa-type territories (Figs. 7.27c and 7.27d).

Very visible is the intensity of the utilitarian interventions – a quarter of them are maintained grass lawns, while a third are vegetable gardens. Clearly, the southeastern part of Dolni Lozen is particularly suitable for the growing vines (Fig. 7.28).

7.3.10   *Pancharevo-Kokalyane*

Just like Kazichene-Krivina, we may consider the settlements of Pancharevo and Kokalyane as a rhizome – not only because of their proximity, but also for their functional connection – commercial and entertainment intertwinings, as well as tangential and shared villa areas. The predominant type of construction in both settlements is modern suburban houses for year-round living, as well as some gated communities. The picturesque nature of the meander Urvich, Lozenska,

Plana and Vitosha mountains, the Zheleznishka and Iskar rivers, Lake Pancharevo, etc. are enriched by strong cultural and historic landmarks.

Pancharevo is located 12 km southeast of Sofia. The village covers 7 km^2, with a population of 2771 people and high density of 379.5 people/km^2. In 1963 it was declared as balneological resort, and some of its central areas were designated for recreation, sports and attractions. Adjacent to the city, high-class villa zones developed, immersed in the lavish and beautiful environment.

Kokalyane is located south of Pancharevo. It has a bigger area of approximately 16 km^2, yet less population (1827 residents), resulting in an almost three times lower population density of 117 people/km^2. At the beginning of the twentieth century a textile factory was built next to the village, and during the 1930s it was converted into a copper factory. Later, the facility was transformed into an irrigation plant for the agricultural territories nearby. The scenic hilly nature has its downsides, too – the villa zones lack running water and some of the permanent residents need to rely on excavated wells to meet their water needs (Fig. 7.29a).

In Pancharevo-Kokalyane rhizome a total of 1483 interventions were mapped: 321 vegetable gardens, 104 greenhouses, 405 vines and vine arbours, 23 orchards, 604 green plots, 20 flower gardens, 6 apiaries (Fig. 7.29b). The adjacent agricultural farm lands were not included (Figs. 7.29c and 7.29d).

There are no significant distribution discrepancies. While 69% of the interventions happen in the Жм zones and 28% in the villa zones, we may conclude, that the various urban agriculture activities are somewhat evenly dispersed throughout the zones. The concentration depends on the scale and the density of the plots – on the intensity of the urban fabric. Not surprisingly, the higher intervention concentration is in the Жм zones. However, there is a slight surprise – the utilitarian aspect prevails in the villa zones on account of the vine plants (Fig. 7.30).

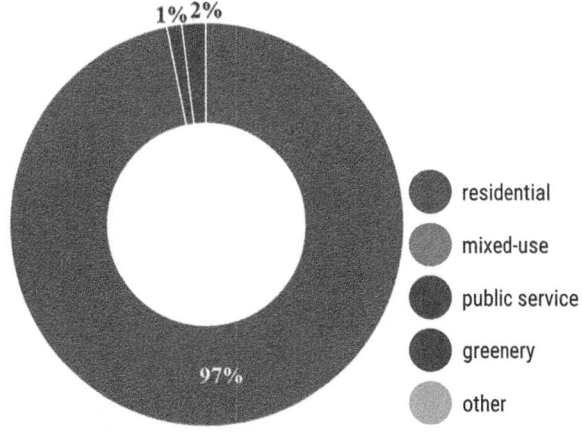

Fig. 7.29a Interventions per zone type

1% 2%

97%

residential

mixed-use

public service

greenery

other

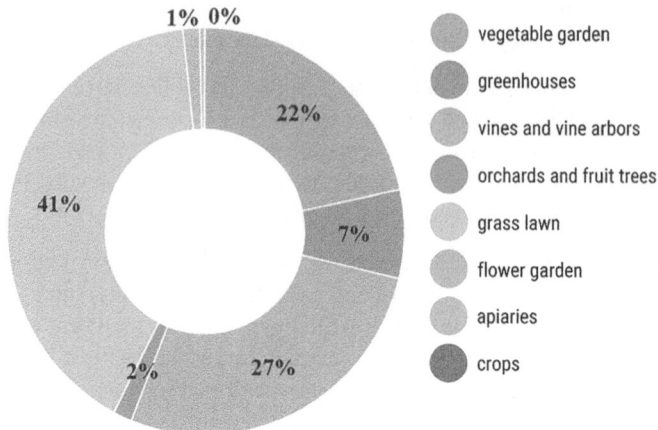

Fig. 7.29b Interventions in Pancharevo-Kokalyane by type

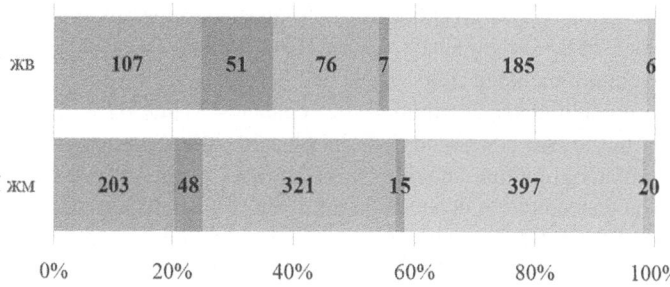

Fig. 7.29c Interventions comparison low-rise resitendial Жм zone and residential villa zone Жв (see Table 7.1)

Fig. 7.29d QR code and link for Pancharevo-Kokalyane map https://cutt.ly/5cSPnvr

Fig. 7.30 (**a**) zones (**b**) all interventions (**c**) vegetable gardens, greenhouses and crops, (**d**) vines, vine arbours and orchards, (**e**) grass lawns, (**f**) flower gardens (**g**) apiaries

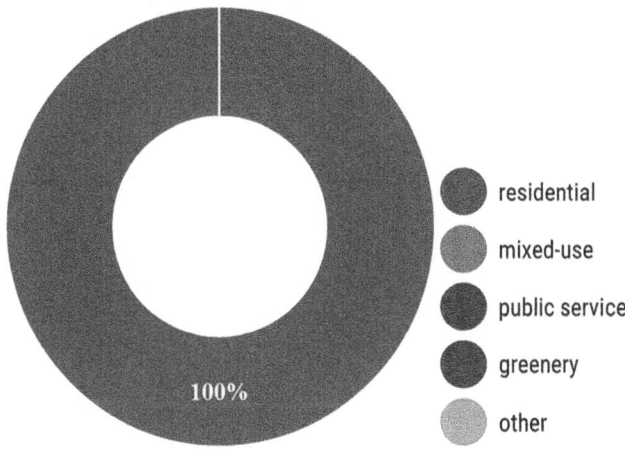

Fig. 7.31a Interventions per zone type

7.3.11   Plana

The village of Plana is specific with its extremely scattered structure. It occupies several small peaks forming separate neighbourhoods connected with narrow curvy roads. Its uniqueness is hidden in the fact that it embraces four mountains: Plana, Vitosha, Verila and Rila. It is famous for the rich pastures in the area. The remote village has only 70 residents, with a density of only 3.8 people/km². The overall area of the Plana lands is approximately 18 km², with an average altitude of 1250 m (Fig. 7.31a).

On the territory of the settlement and part of its adjacent zones are registered 233 interventions, distributed by type as follows: 66 vegetable gardens, 41 greenhouses, 30 vine arbours, 16 orchards, 80 garden plots (Fig. 7.31b). All of them occur in residential zones, and 82% in the central low-rise Жм residential zone. A significant difference in the urban agriculture practices are observed between the types of residential territories (Figs. 7.31c and 7.31d).

Plana is surrounded by forests, vast pastures and seemingly endless agricultural lands. However, according to the General Development Plan, some of these territories would be reassigned from agricultural to residential and even to mixed-use. Like Dolni Pasarel, the proportion between vegetable gardens and the greenhouses – approximately 2:3 – is striking. Another discrepancy with the previously observed results in comparison with the other settlements is the small amount of vines and

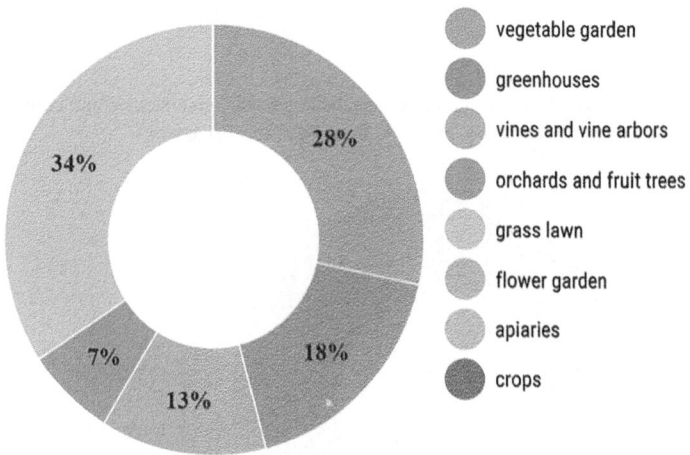

Fig. 7.31b Interventions in Plana by type

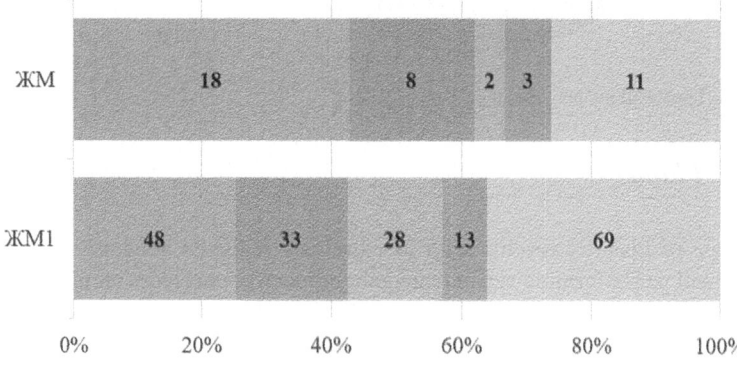

Fig. 7.31c Interventions comparison between low-rise residential with restrictions Жм1 zone and low-rise residential Жм zone (see Table 7.1)

Fig. 7.31d QR code and
link for Plana map
https://cutt.ly/wcSPS6J

Fig. 7.32 (**a**) zones (**b**) all interventions (**c**) vegetable gardens, greenhouses and crops, (**d**) vines,
vine arbours and orchards, (**e**) grass lawns.

vine arbours – only 13%. These variations could be explained with low temperatures
and the harsh mountainous climate (Fig. 7.32).

7.3.12 Analysis

During the field research (2017–2020 period), overall 13,029 urban agriculture
locations were identified and mapped in their real GPS coordinates. The documented
interventions do not claim to be exhaustive (as they usually vary on a seasonal and
annual basis), but rather to give a snapshot of the dynamic bottom-up urban
agriculture practices.

On an individual level some interesting observations were made:

The practices of urban agriculture in the residential complexes and central urban areas often go beyond the pure decorative aspect and vegetable gardens are emerging, sometimes hidden from the view of passers-by. A very surprising inversion was spotted concerning the exact location of the vegetable gardens. In the low-rise residential areas, the gardens are situated mostly in the backyard of the individual plots, hidden from plain sight, away from the street traffic and pollution, while receiving a good amount of sunshine (Fig. 7.33b). In the high-rise residential areas, the vegetable gardens are usually in the periphery, hidden amidst the dense trees in the public greenery, or behind some free-standing garages. However, besides the desire for cleaner space and suitable sunshine, there is another underlying aspect for hiding these agriculture interventions. Making them difficult to spot is for security reasons as well – according to "Ordinance for construction, maintenance and protection of the green system of Sofia Municipality", Art. 36 (2) 11, it is forbidden to gather seeds, fruits, cuttings, and to pick herbs. The same phenomenon was observed in Triaditsa (Fig. 7.33a). Similarly, the apiaries are also located near the border of the district, away from intense urban life. However, with an ordinance from 2020, the apiaries are equated to livestock farms and are subjected to a number of regulations (Fig. 7.33b).

Exploring the maintained grass lawns, two types of interventions were identified. There is often a clear distinction – for representative and for private purposes. An indication of this was the observed duplication of the representative lawn in front of the house, as the secondary lawn is located in the heart of the property, in direct connection with the vegetable garden and the orchard, thus becoming a more intimate space for the household (Fig. 7.34a). In other cases, the vegetable and

Fig. 7.33a Hidden vegetable gardens in Triaditsa District

Fig. 7.33b Hidden vegetable gardens in Mladost region

Fig. 7.34a Grass lawns in Gorublyane

fruit plantations or vine arbours are completely replaced by extensive lawns. These cases are most common in the outskirts of the smaller settlements, in newly built properties and/or in villa zones (Fig. 7.34b).

In areas with low-rise housing or in smaller rural settlements, the interventions are a result of personal and domestic behaviour, and the primary beneficent is the

Fig. 7.34b Grass lawns in Bistritsa villa zone

household itself. Nevertheless, there are some indirect benefits for society at a community level.

The size of the front lawn may vary between the villages, in accordance with the altitude, the urban street network, the sun direction, etc. However, the role of the front lawn expands way beyond the boundaries of the property and the household (Fig. 7.35a). Basically, it aestheticizes the immediate public environment and creates pleasant street silhouettes. Even when the fence is high and opaque, there are often small flower beds or tiny grass patches that beautify the sidewalk and speak about the owners' preferences and gardening skills (Fig. 7.35b).

Another type of very striking benefit for the community was traced within a small detailed side-study performed in Mladost residential complex. It focused on the continuing tradition and the central role of the vine arbour. Both in the public spaces and in the private yards, the vine arbour is an essential spatial and visual benchmark, ensuring a delicate balance between decorative and utilitarian functions. It keeps alive the historical practice of creating shared multifunctional spaces – a smooth transition between interior and exterior, housing many different outdoor activities, providing a pleasant environment for recreation and communication (especially during the hot summer) and providing an opportunity to enjoy the sweetness and the aroma of its fruits (Fig 7.37a). In Mladost 1–4 regions most of the vine plants were situated according to the suitable sun exposure (Fig. 7.36a), even though sometimes it happens to be on the back side of the apartment building. On the other hand, the vine arbours were mostly located in front of the building entrances, often accompanied by do-it-yourself benches, tables and gazebos, turning the space into a co-shared exterior living room for the residents – coffee time for elderly ladies, snacks and drinks for the gentlemen, gatherings for the youngsters. Often another

Fig. 7.35a House in Bistritsa

Fig. 7.35b Street silhouette in Lozen

phenomenon can be observed – the vine plants reach up quite a few floors in height, providing much desired sunshade for the upper floor balconies during the hot summer (Fig. 7.36b). All these applications of the simple vine serve as a strong

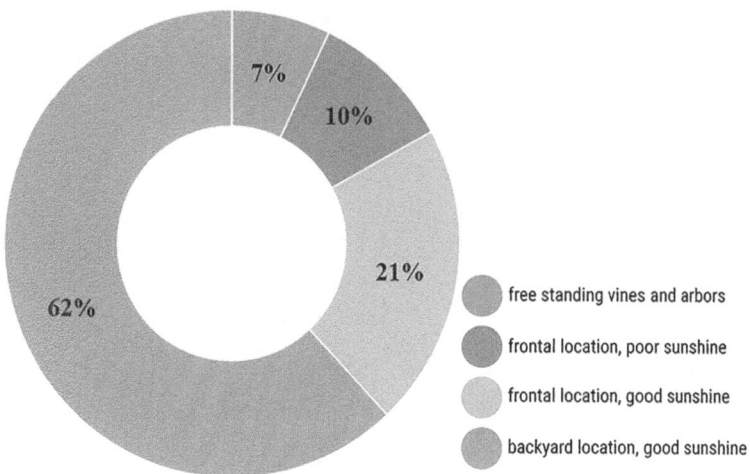

Fig. 7.36a Orientation of the vines and the vine arbors in accordance both to the main entrance of the residential building (free standing, frontal or backyard) and the quality of the sun exposure

Fig. 7.36b Location of the vines in accordance to the building and the cardinal directions

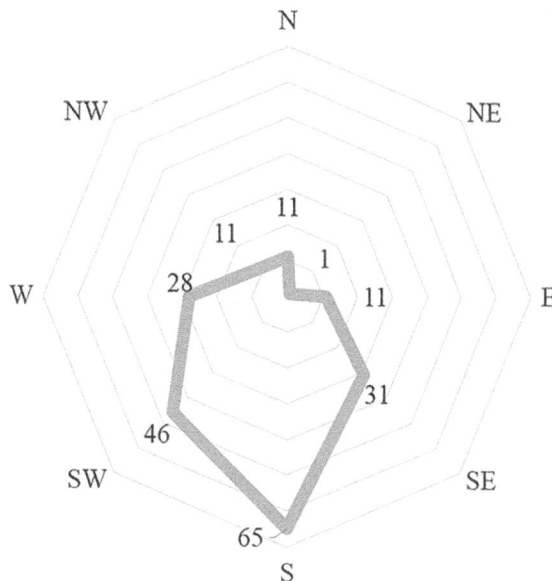

demonstration of the deep-rooted utilitarian urban agriculture tradition (Figs. 7.37a and 7.37b).

On a societal and urban scale, the practices of urban agriculture are more or less a result of neighbourly relations and interactions. Urban agriculture practices have a synergy effect – the active lifestyle and the amount of bottom-up interventions in the environment often serve as a fine example to follow and thus stimulate other

Fig. 7.37a Entrance to an apartment building with lavish vines and vine arbors and the shades they provide

Fig. 7.37b Multiple vines climbing the facade of a residential building to the upper floors, providing shade and grapes for the residents

participants to join in and to contribute to many other spatial interventions in the surrounding areas, creating new and maintaining existing lawns and flower gardens, constructing new vine arbours, gazebos, outdoor gathering places, etc. This not only enriches the environment, but stimulates societal cohesion by enjoying together the co-created outdoor spaces for shared recreation (gathering for a chat, consumption of food and beverages, board games in front of the block, construction of urban furniture and barbecues, children's facilities, etc.) and thus fosters responsible communal behaviour (Figs. 7.38a and 7.38b).

Fig. 7.38a Vine arbours in a family house

Fig. 7.38b Shared space in front of an apartment building

On an urban scale, two main aspects of urban agriculture can be distinguished – decorative and utilitarian – by introducing a utilitarian-decorative coefficient. As the vines traditionally have both a utilitarian and decorative aspect, the coefficient can be determined as follow:

$$UDC = \frac{Nveg + Ngreenhouses + Nvines + Norchards + Napiaries}{Nvines + Nflowers + Ngrasslawns}$$

This helps us to transcribe the following charts for Triaditsa, Mladost, Gorublyane and Pancharevo District (as a whole). Impressive is the shift of the UDC in accordance with the change with the urban environment specifics (a particularly striking difference between the residential areas (Mladost, 1, 1A, 2, 3, 4) and sub-urban areas with single-family houses and yards). This major shift from the decorative towards the utilitarian aspect of the various urban agriculture practices corresponds to the shift from the central-urban towards the peri-urban and rural environment and their different urban usage and lifestyles (Fig. 7.39).

A similar analysis for the settlements in Pancharevo District shows clearly the nature of the lifestyle in the different settlements (Fig. 7.40). There is a significant contrast between Pancharevo-Kokalyane rhizome and Bistritsa (respectively with UDC = 0.83 and UDC = 0.86) and the Kazichene-Krivina rhizome and Dolni Pasarel (respectively UDC = 2.27 and UDC = 2.97, which signifies the predominantly utilitarian aspect of the urban agriculture practices).

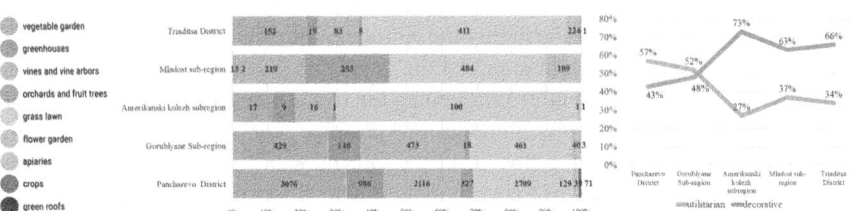

Fig. 7.39 Analysing the UCD for the three districts and two sub-regions

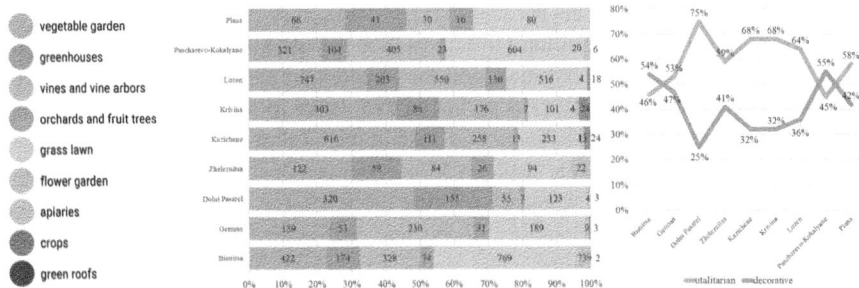

Fig. 7.40 Analysing the UCD for each settlement in Pancharevo District

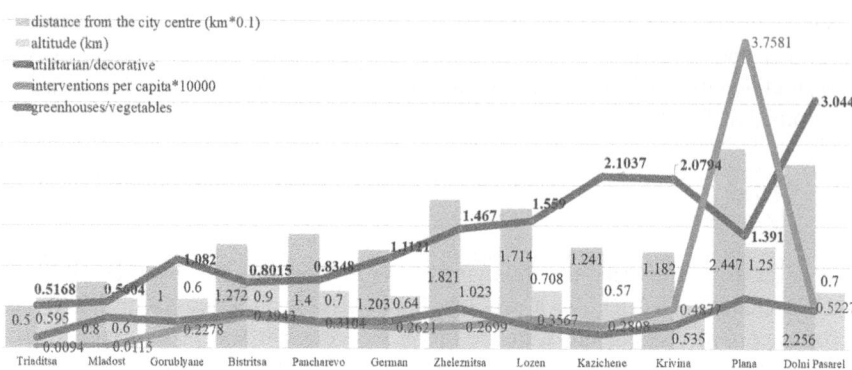

Fig. 7.41 Comparison between several aspects: distance from the city centre, altitude, proportion utilitarian vs decorative interventions, coefficient of interventions per capita and proportion of greenhouses per vegetable gardens

As expected, in the central-urban territories, or in zones with mid- and high-rise residential buildings, the utilitarian aspect is existent but not most important, while in the rural territories or in zones with low-rise residential building, it represents a significant part of the urban agriculture practices. During the secondary analysis some interesting correlations occurred (Fig. 7.41). The further the village is situated from the city centre the higher is the number of interventions per capita. Also, the further away the village is located from the city centre the higher the utilitarian-decorative coefficient. Another interesting, yet not surprising correlation is observed with the change of altitude – the higher the altitude of the village, the higher the ratio coefficient between the greenhouses and the vegetable gardens. A simple explanation for this is the natural desire for shortening the food chains in the remote areas by using more and more various home-grown produce (for logistic and economic reasons). A fine example for this is Plana village with its altitude of 1250 m, dispersed along some peaks of Plana mountain, which has the highest intervention intensity per capita.

Another confirmation for the above-mentioned observations can be found in the comparison between the various agricultural practices per capita, compared to the population density and the distance from Sofia – the further the village is from the capital, the more interventions per capita (Fig. 7.42).

At a city level, urban agriculture improves the land use both within the individual properties, the mixed-use territories, and the abandoned and desolate public areas. We may perceive these practices as an instinctive desire to reintroduce nature in the urban environment and thus actively and sustainably counteract the deterioration of the urban environment and the negative consequences of the intensifying building investment intentions. Yet, when we overlay the territories for future urban development over agricultural territories, forest, etc, we realize that the situation will become only more challenging. The major expansions of residential areas are planned not only around currently overpopulated areas (Gorublyane, Kazichene and Bistritsa), but also around settlements with an extremely low population density,

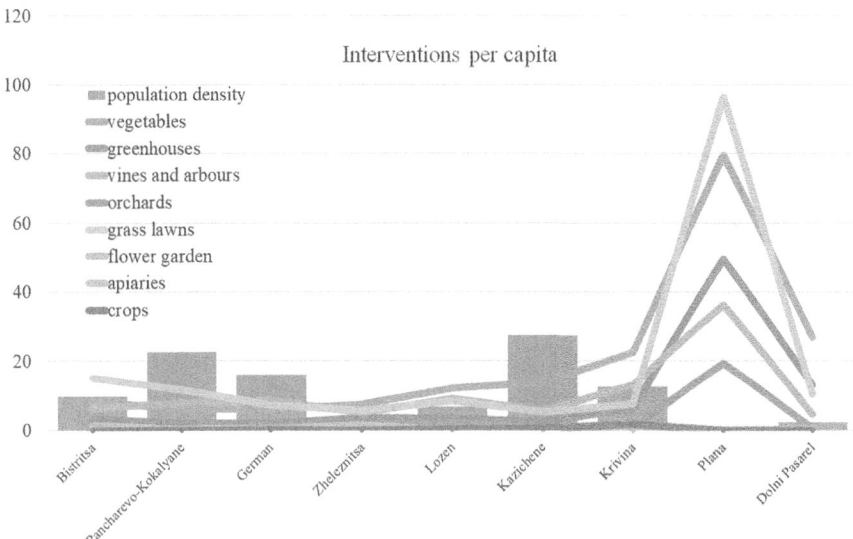

Fig. 7.42 Intervention types per capita compared with the density of the settlement and the distance from the Sofia

with many abandoned lots (like Plana, Dolni Pasarel and Zheleznitsa), threatening to deepen the negative consequences of overpopulation, sealing of agricultural lands, pastures, forest and inducing more pollution in the otherwise clean mountainous natural environment. Similar processes occur within the intra-urban and the peri-urban districts of Triaditsa and Mladost. The reduction of the park areas, the closing of the green wedges, the discontinuity in the green belt due to repossession of state-owned properties threaten to destroy the proper function of the green system implanted in the urban fabric in order to serve as a supplier of clean air from the mountains and to counteract the air pollution from overpopulation and the intense traffic (Fig. 7.43).

7.4   Lessons learned

We may say that urban agriculture practices improve the interpersonal relationships and strengthen the fragile connection between people, the city and the natural environment. Urban farming methods give us a good opportunity to create a favourable setting that stimulates more outdoor activities, long walks and the use of alternative ways of ecological transport.

On an individual level, engaging in some form of urban agriculture encourages a more active lifestyle, creates a pleasant share of personal heavenly peace. Even nowadays, the functional organization of the domestic yard still follows the

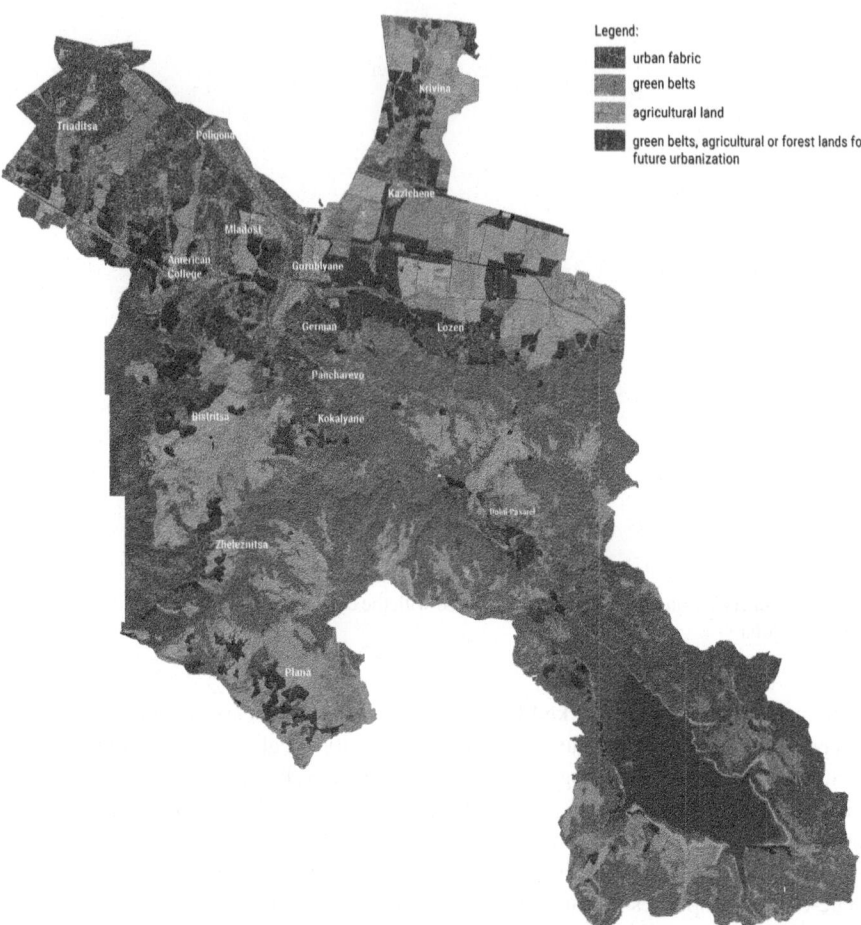

Fig. 7.43 Tracing territories designated for construction which were previously green or agricultural lands – the red sites are designated for future urbanization and construction

evolutionary path set by the centuries-old practices. The vine arbours (or other large fruit trees), ennoble the yard, provide proper shading and become a major functional and spatial benchmark – a shared outdoor workspace for household members. The produce from the garden becomes not only essential, but even preferred not only for its economic value, but also for its better qualities and for the pure satisfaction of the effort that has been put into it.

On a community level, the individual environmental interventions bring about a strong synergy effect. The active lifestyle and the accumulation of good practices set a good example to be followed and in turn stimulates the neighbours and the other members of the community. The bottom-up spatial development (nurtured grass lawns, colourful flower beds, rich arbours and shared spaces) activates a complete renewal and aestheticization of the environment and creating pleasant, neat and clean street silhouettes, which could be expected to enhance satisfaction with the immediate urban environment.

On a city level, we may say that urban agriculture practices lead not only to reclaiming existing abandoned spaces and green areas, not only to re-cultivating the existing zones, but to development of new ones, to improving the vision of green spaces and a clean urban environment, thus increasing air quality and improving the quality of life while preserving the natural environmental conditions in cities.

On the other hand, the wide variety of urban agriculture practices creates a unique functional and spatial pattern, which greatly contributes to the biodiversity and to the uniqueness of each settlement, and which utilizes the natural resources in an optimal way (Fig. 7.44).

In conclusion, all these bottom-up processes could and should be supported by appropriate regulations and planning policies in order to cultivate various specific aspects of urban identity, and not only to upgrade the existing historical traditions, but to become an ambassador of the principles of sustainable development in all stages of architectural design, urban planning and territorial development in Bulgaria.

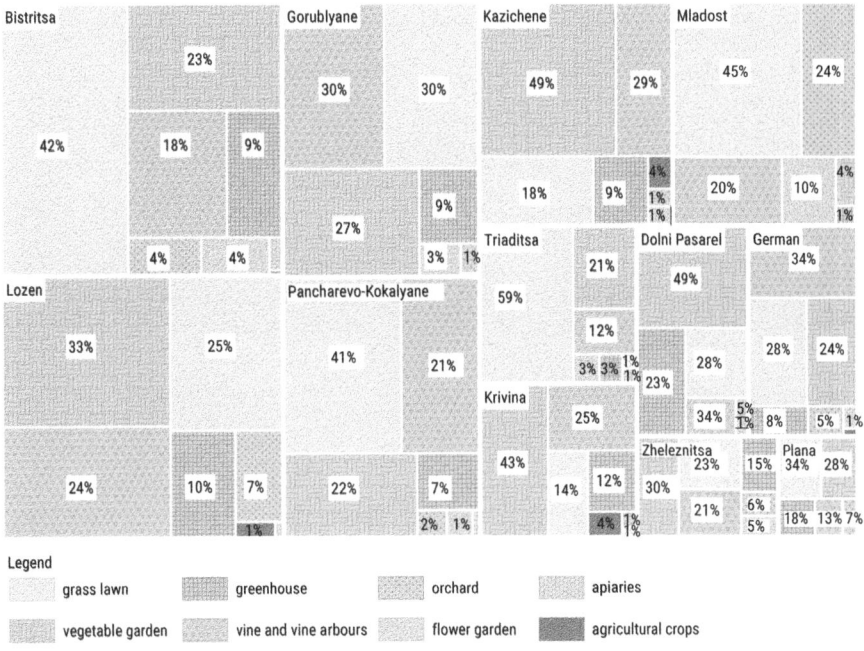

Fig. 7.44 Combined analysis of number and proportions of interventions in all districts and settlements

References

Assolombarda. (2015). Expo 2015 Feeding the planet. Energy for life. Retrieved February 16, 2021, from https://web.archive.org/web/20130927085711/http://www.expo2015.assolombarda.it/tema

Dimov, H. (1976). *Renaissance Yards in Kotlensko*. Zemizdat (In Bulgarian).

Howard, E. (2010). *To-morrow: A peaceful path to real reform*. Cambridge University Press.

Howard, E. (2013). *Garden cities of to-morrow*. Routledge.

Matera Basilicata Foundation. (2019). Gardentopia. Retrieved February 2021, from https://www.matera-basilicata2019.it/en/programme-2019/themes/utopias-and-dystopias/1411-gardentopia.html

Puhalev, B. (1984). *The first gardens of Plovdiv*. PhD thesis (In Bulgarian).

Robev, R. (1981). *Yard gardens in Koprivshtitsa*. Zemizdat (In Bulgarian).

Sugarev, D. (1956). *Karlovo Yards during the Renaissance*. Bulgarian Academy of Sciences (In Bulgarian).

Toleva-Nowak, N. (2019). *The Urban domestic Yard in the Bulgarian revival – A prototype of Urban agriculture as a strategy for improving the quality of life of Urban communities*. Scientific Work of the Union of Scientists Plovdiv, Series B. Natural Sciences and the Humanities, (19), 234–242.

Chapter 8
Conclusions

Dona Pickard

The results from the study "Urban agriculture as a strategy for improving the quality of life of urban communities" support the assertion in the literature that UA can alleviate pressing urban problems. The findings of the book demonstrate that people, motivated by the search for good quality food, physical and emotional well-being, environmental responsibility, social connections or just pleasurable leisure time, can re-think and improve their quality of life, which can also reflect positively on the quality of life of their communities and the livability of the city as a whole. Overall, we presented data to support the claim that UA could be implemented locally to tackle social inequalities, poverty, social cohesion and unemployment, through environmentally friendly and fair use of public space. The research questions we posed at the beginning of the research were if, how and to what extent these positive effects of UA manifest themselves, and how they can be harnessed to the best interest of individuals, the local communities and cities in general so that the quality of life for citizens is improved.

The results from the reviewed literature on the topic, together with the presented empirical data from the research and analyses suggest that while UA activities could potentially deliver a wide range of benefits for individuals, households and city communities, as well as for the urban environment, not all positive UA effects are possible or imminent in every location where UA is implemented. Our research provided evidence that for UA to be able to improve any aspect of urban QoL, it is crucial that the right local conditions for it should be in place. These include not only the available natural resources for agriculture, such as land, water and favourable climate conditions, but also social, political and cultural factors that determine social capital, collective action and public attitudes and values related to agriculture, the environment and city space. Therefore, the capacity of UA to improve the QoL of urban communities is not an inherent characteristic of UA activities themselves, but

D. Pickard (✉)
Institute of Philosophy and Sociology, Bulgarian Academy of Sciences, Sofia, Bulgaria

© The Author(s), under exclusive license to Springer Nature Switzerland AG 2022   185
D. Pickard (ed.), *Urban Agriculture for Improving the Quality of Life*, Urban Agriculture, https://doi.org/10.1007/978-3-030-94743-9_8

it depends on how the local resources for UA are managed. This, in turn, is a complex function of the interaction between local authorities, public institutions, civil society and individual actors. Our findings testify that UA cannot be just "plugged in" the social, economic and environmental urban systems and their spatial structure, and be expected to become an inseparable part of urban life which brings about positive trends in the life of urban communities. It is the processes of initiating, organizing, governing and sustaining UA practices that can make UA successful in raising urban QoL. As such, UA as a successful strategy for improving the quality of life of urban communities is very much socially, culturally and politically determined. As the lessons learned from our multidisciplinary studies have demonstrated, it is all about the interplay between all factors and relevant actors that make UA successful (or not) in these aspects.

Literature has shown that in some places UA is driven by crises and need, and when in these locations adequate policies and regulations have followed, UA practices could lead to improving the well-being of citizens. In other places, policies are deliberately developed to spearhead expected positive effects and this has also proved a viable way to mobilize the local resources and make the best use of newly initiated practices. In other places, traditional practices and modern approaches co-exist and policies adapt to this co-existence. In all cases the success of UA as a way to improve QoL depends on the effective interaction between all stakeholders. It is because of this interconnectedness that a higher level of governing approach is needed to ensure the local UA potential to improve citizens' well-being is used to its fullest. A concerted effort by all parties and stakeholders relevant to UA and its beneficial effects would allow a comprehensive strategy for UA to unfold and bring about positive change that flourishes on socio-economic, cultural, environmental and spatial synergy, rather than for only single and limited benefits of UA to emerge sporadically and for a restricted number of locations or privileged communities.

In the case of Sofia, the research shows that where there is no formal institutional recognition of UA and it is left out of the control and regulations of the authorities, there is little that UA can contribute to the well-being of a broader range of urban communities. Just because traditional individual forms of UA exist and are well rooted in a location, this would not necessarily help improve the QoL of urban communities in general. When the latter is a goal, city level policies are needed, as small, private and individual UA initiatives alone do not offer the potential for significant improvement of city well-being. This is especially true when the challenges at city level are such that no one particular group or type of stakeholder could solve it by themselves. These include pollution of potential UA sites that needs to be mitigated, pressure for construction on open space and farmland, barriers to cooperation between small and traditional farmers, low education status and lack of social, entrepreneurial and other "soft" skills of potential UA target groups, lack of finance and lack of supportive policies. Coordinated actions at the individual or business level, as well as at civil and institutional levels are needed in order for these deficits to be overcome. These actions could be organized as part of large civil campaigns or channeled through local policies, but in either case, they should be also based on a thorough understanding of the motivations of practitioners and the needs of the targeted local urban communities.

Methodologically, we approached the otherwise closely interlinked aspects of the concept of quality of life, from four disciplinary viewpoints – social, economic, environmental and spatial. The use of different scientific methods allowed for a more in-depth study of the various mechanisms through which UA can impact on different spheres of the life of individuals, communities and the city environment, depending on their nature. The multi-disciplinary approach also allowed us to make a distinction between real positive outcomes of UA and its potential benefits. This distinction needs to be kept in mind when researching urban agriculture, which is all too easily hailed as a champion of environmental and social innovation in cities but it is often difficult to distinguish how exactly UA brings about change, for whom and at what price.

Still, the potential benefits of UA are of special research interest as commonly the factors that have kept that potential from being realized in practical outcomes reflect the local social, economic and cultural specificities. As already demonstrated, they are at the crux of UA success as a tool for improving the QoL of urban communities. Linked with this is another methodological issue – the difference between studying the effect of UA on practising individuals and communities compared to studying the effect of UA on citizens who are not directly involved in gardening activities. Our experience of designing and conducting the research shows that, when investigating the effects of UA on urban life as a whole, it is very important to be vigilant with the indicators used and to resist the temptation to extrapolate the results found among practising gardeners on to the general community. While it is rather straightforward to design research instruments to study the impact of UA on the QoL of practising urban gardeners, researching how UA influences the QoL of members of non-practising communities entails more elaborate work that evaluates this influence through a combination of methods. These methods should be chosen carefully so that they can enlighten the processes of UA changing the environment in which these communities live and interact and which, in turn, can lead to a change in the QoL of the non-UA-practising citizens.

To sum up, our study focused on the mechanisms that stand behind urban agriculture's positive impact on the social and economic well-being of urban communities as well as on the more fair and sustainable use of urban space and the urban natural environment. The results from the study revealed that the links between UA and its benefits are not straightforward, and they depend on a range of local factors. Also, in different cases and locations, various types and elements of UA can yield different results. We presented data from Sofia on what effects UA can have and has shown to have at three levels – the individual and household level, the community level and the city level.

At the individual level, the primary benefit of UA is the provision of food which is trusted by consumers to be of good quality and produced in environmentally friendly ways. We also found evidence that UA contributes to feeling physically and emotionally fit, mostly due to the physical activity it is associated with, as well as the satisfaction with the healthier diets it can provide. The self-reported emotional well-being is significantly more common than the physical health, reported by respondents. While respondents from lower income categories seem to be slightly more likely to engage in UA in Sofia, our research has not supported the claim that UA significantly improves the quality of life of disadvantaged social groups. We have demonstrated that because of structural barriers to integrating communities through UA, their well-being cannot be improved without targeted cross-sectoral policies at local level. It must be noted that while UA can benefit physical and mental health, our study suggests it is likely that physical and mental well-being might be a pre-condition to get involved in UA in the first place, which – once again, points to the need for targeted policies that could encourage individuals with special physical or mental needs to take up UA.

Another significant effect of UA is the feeling of empowerment and ability to exert control over one's life, especially for the elderly who garden together with people from younger generations. This effect works on the basis of the need of the elderly to feel recognized, important and appreciated, if they do not have any other responsibilities, such as dependent children or expectations at their workplace. For children and adolescents in kindergartens and school UA helps to develop important social skills such as improved work ethic, responsibility, skills for joint activities, teamwork and mutual help. These, together with raised environmental awareness are closely linked to enhanced inner QoL chances for a better life.

From an economic perspective, it is easy to discern the effects of UA at individual and farm level. For commercial initiatives that sell their produce in the same settlement, the advantage of being close to their market is a logical consequence. For non-commercial practitioners, UA also makes economic sense as long as the produce is consumed in the household and it represents savings that could be spent on other items. The consumption of healthy food is also thought to improve one's health and thus help citizens save money on healthcare due to the healthier lifestyle.

At the community level, our research suggests that UA can contribute significantly to communities' social cohesion, social inclusion of disadvantaged groups and to civil participation when UA practices are well-established in the particular location and the local population is aware of their existence and the options for participation that they offer. Such a local environment allows for UA to foster a stronger democratic culture, an improved social environment and more active social engagement of people towards improved QoL of their communities. Without this scale of UA development, it is difficult to improve the quality of social interactions at community and city level and similarly, it is too difficult to transform the social space and make it more inclusive. On a smaller scale, what is possible and has been shown to produce well-accepted results, is the aestheticization of abandoned plots, even if no community identity has been developed around these plots.

At the city level, the most widely accepted positive effects of UA are related to the ecosystem services that it offers. These include cutting CO_2 emissions, reduction of run-off water and the general improvement of the microclimate. Again, for these effects to be felt by the citizens, UA needs to have spread on a wide scale, so that a network of green productive spaces is in a strong position to mitigate the environmental pressures on cities' air, water and soil. UA that is spread so widely can also boost the local economy by serving as a basis for UA support businesses and also by contributing to the city budget, given the high growth rates of commercial urban agriculture initiatives. In the long run, and with structured educational support for the youngest generations in terms of food awareness and more sustainable consumption patterns, a culture of sustainable living could be spread more widely.

Additionally, UA practised individually or as a food business and the newly created marketplaces and farmer's markets related to it in and around cities, create alternative food supply chains and networks, improves the accessibility of good quality produce and enriches the food-related choices that citizens have.

Overall, our study demonstrates that UA can have significant positive effects on the QoL of urban communities and therefore it is worth supporting at a local level. There are three main reasons to promote UA that have been demonstrated in the book.

Firstly, when the potential of UA is fully utilized, the benefits of urban gardening practices are not limited only to the individuals and communities of comfortable social and economic status who are civilly empowered and enjoy a satisfactory QoL level. The full potential of UA allows for it to mitigate the risks of poverty and social exclusion and bring positive change in the material and physical well-being of less advantaged social groups, improve their relationships with other people and their social capital, as well as their overall satisfaction with life. If UA is not recognized as a tool for supporting disadvantaged social groups by the local authorities, its potential for improving their well-being would be squandered.

Secondly, UA can help mitigate the adverse effects of climate change and potential biological crises such as global pandemics, which already affect cities around the globe. As these threats are becoming a daily part of urban life, the food chains of big cities are put under stress and their security is threatened. As public authorities bear the responsibility for public health, it is a forward-looking strategy to support UA in order to improve the sustainability of urban food systems.

Thirdly, in cases like Sofia, where the general public is not aware of the possibilities that UA offers for an improved QoL, it is important that the local authorities or other civil actors take action to raise awareness of the potential of UA to make people's lives healthier, more financially and food secure, as well as more inclusive, environmentally-friendly and pleasant. While UA's public image is largely a result of cultural and historic circumstances, it is also dependent on current public discourse and institutional principles that shape agriculture in the urban environment in a more or less favourable way. As our empirical data have demonstrated, if the authorities express disregard of the phenomenon's social, economic, environmental and cultural significance, the public mirrors this attitude and perceives UA merely as an agricultural undertaking in the traditional sense.

In the course of our research we identified several issues to be taken into account when urban policies are designed for supporting UA, which can be formulated as policy recommendations. They would ensure that the maximum of UA's potential to improve the QoL of urban communities is utilized and are valid for all types of local circumstances:

- Urban agriculture produces the highest impact on the broadest range of social groups, communities and public spaces when it is normatively regulated and the regulations explicitly recognize its specific forms and effects (educational, social, health-related, economic, environmental) on specific social target groups and

elements of the environment (children, the elderly, small businesses, disadvantaged groups, protected species or natural zones, etc.).

- All evidence shows that the most effective UA governance happens not through
  one strategy, regulation or directive, but through integrating UA practices in all
  sectors of governance that are related to it. It is through cross-sectoral measures
  that the multi-faceted character of UA can be utilized in the best interests of all
  target groups to which it is geared.

- Regulatory forms of support are generally more effective than financial support
  for long-term UA development. While financial support is commonly channeled
  into time-limited projects that find it hard to sustain their activities after the
  financial help runs out, normative and regulatory support strengthen the deep
  foundations for UA. These include encouragement of local food consumption,
  including in food procurement, limiting soil sealing, especially in areas with
  agricultural traditions and good agricultural quality soils, and protecting UA as
  cultural heritage practices;

- To have efficient UA policies, they need to be based on testable data and evidence
  using measurable indicators for evaluating UA's effectiveness for tackling urban
  challenges. Depending on the goals and priorities of urban communities and local
  authorities, these data and indicators could be either narrower (focusing on social,
  economic, territorial and environmental issues), or more encompassing and
  offering information on the whole food system and urban space use and
  governance.

- Community and city-level decision-making processes regarding the development
  of UA should involve all stakeholders. The constant dialogue between civil,
  public and business actors as well as all relevant structures of the local governance is a prerequisite for the sustainable and adequate development of UA. The
  interaction between all stakeholders is a key to the success of individual UA
  practices, as well as to the visibility of the local authorities' support for UA,
  which often remains unnoticed and underused by the larger public. None of the
  actors that are relevant in the development of UA could effectively use UA to
  improve the QoL of individuals and communities without close cooperation with
  the others. Burdening only one group of actors with responsibilities to work
  towards this goal makes UA initiatives less socially and financially sustainable
  and decreases their capacity to adequately address the wide range of urban
  problems that could be alleviated through urban agriculture.

To a large extent, the results of the study "Urban agriculture as a strategy for
improving the quality of life of urban communities" are derived from the case of
Sofia – the capital of Bulgaria. Still, they outline the main aspects in which urban
agriculture practices can have a positive impact on the socio-economic well-being of
people and communities in cities in general, on their environment, as well as on the
livability of their urban space. While UA practices usually have a particular focus or
intended function, they can all have a wide range of intended or unintended
secondary effects: on social relations, life satisfaction and self-esteem of people,

maintaining sustainable niche economic activities, creating harmonious and inclusive public spaces and contributing to the conservation of biodiversity in the city.

Based on the results of the study, we can confidently say that UA has a multidimensional impact on the quality of life of urban communities. Its effects on the QoL at the individual, community and city levels are not equally strong and direct, but they all allow for improvement of the social well-being, economic opportunities, environmental quality and public spaces of cities. To what extent it would contribute to the different aspects of the urban QoL depends largely on the dialogue and cooperation between the relevant local actors – civil and public organizations, businesses and the local authorities. While there are multiple driving forces behind UA, our study demonstrates that the collective effort of all relevant actors and stakeholders is required to transform it into an urban phenomenon with significant role in improving the quality of life of urban communities.

Appendix

Self-Assessment Tool to Measure the Potential of a UA Initiative to Improve the Quality of Life of Urban Communities

To be useful to UA practitioners, we have devised a self-assessment tool to help individuals, organizations and local authorities estimate to what extent an urban agriculture practice or initiative has positive effects on the social, economic, environmental and spatial aspects of urban QoL, and which of these effects are most typical of it. It is a tool to assess the UA contribution to improving the quality of life in cities, based on the project's empirical outputs and adapted indicators from the European COST action UAE Europe. The COST action study served as a conceptual ground to develop the self-assessment tool, and the indicators and values of UA's societal benefits that were developed by it are applied here as well. However, the self-assessment tool we devised is structured around the social, economic, environmental and spatial aspects of urban QoL that we studied in Sofia. The tool was tested in three UA cases in Sofia – an educational garden in a state school, a community garden and an experimental field of a research institution. The testing demonstrated that while the main function of the UA initiative was reflected in the results, the tool also allows the practitioners to recognize unintended positive effects of the initiative that they were unaware of before. Thus, practitioners can utilize these effects by developing them purposefully, monitoring them more closely and deliberately promoting them.

The tool comprises of four sets of questions that refer to the social, economic, environmental and spatial aspects of urban QoL. It includes instructions on how to calculate your points for each aspect of the elements of quality of life and how to interpret your overall result by marking your scores on the radar graph at the end. Some issues come up in more than one aspect, due to the cross-sectoral impacts of UA.

© The Author(s), under exclusive license to Springer Nature Switzerland AG 2022   193
D. Pickard (ed.), *Urban Agriculture for Improving the Quality of Life*, Urban Agriculture, https://doi.org/10.1007/978-3-030-94743-9

Social Aspects

For each Yes answer, please mark 1 point in the blank space

1. Do participants in the initiative represent various:

   Professions_ _ _
   Age groups_ _ _
   Social status groups_ _ _

2. Has the initiative resulted in new acquaintances, leading to contacts among its members in other areas of everyday life (e.g. celebrating personal holidays, travel, participation in other civic initiatives)?_ _ _
3. Have the participants in the initiative created public goods that can also be used by non-participants (such as free-access benches, playgrounds for children)?_ _ _
4. Does the initiative involve activities to support disadvantaged children or the elderly (e.g. therapy for disabled children, provision of soup kitchen products, etc.)_ _ _
5. Does the initiative involve targeted educational activities (e.g. inclusion of gardening in biology, agriculture classes or as an educational activity for schools, kindergartens, qualification or retraining, etc.)?_ _ _
6. Are representatives of the local/neighbourhood community involved in decision-making on the initiative development?_ _ _
7. Does gardening involve the regular interaction between more than one type of social actor (individual citizens and families, civic associations, educational institutions, non-governmental organizations, businesses, consultants, local and central government bodies)?_ _ _
8. Does gardening put a special emphasis on protection of traditional varieties/breeds, local agro-technical practices, etc.?_ _ _

RESULT FOR THE SOCIAL INDICATORS

Total number of points:_ _ _

Economic Aspects

For each Yes answer, please mark 1 point in the blank space, and for the scale-based answers, please mark the points corresponding to your answer.

9. What part of the produce generates income (basic or additional) for members of the garden through sale? _ _ _

0%	0–25%	26–50%	51–75%	76–100%
0 pts	1 pt	2 pts	3 pts	4 pts

10. If part of the produce is placed on the market, what part of the sales is generated directly between producers and end users?

0%	0–25%	26–50%	51–75%	76–100%
0 pts	1 pts	2 pts	3 pts	4 pts

11. How many paid jobs does the initiative provide (equivalent to full-time jobs)?

0	<0.25	0.25–1	1–5	>5
0 pts	1 pts	2 pts	3 pts	4 pts

12. What part of the produce of the garden/initiative is sold as processed goods?

0%	0–25%	26–50%	51–75%	76–100%
0 pts	1 pts	2 pts	3 pts	4 pts

13. What is the total value (even if not sold) of the produce from the garden/initiative on an annual basis in EUR or USD?

<1000	1000–10000	10000–25000	25000–50000	>50000
0 pts	1 pts	2 pts	3 pts	4 pts

14. Is the produce or part of it branded/marketed as "grown near the city/local"?_ _ _
15. Is the produce or part of it branded as offering local/traditional varieties/breeds/products?_ _ _
16. Is the produce or part of it branded/marketed as "gourmet/boutique"?_ _ _
17. Is the produce or part of it branded/marketed as "clean/organic/environmentally sustainable"?_ _ _

RESULT FOR THE ECONOMIC INDICATORS

Total number of points:___

Spatial Aspects

For each Yes answer, please mark 1 point in the blank space, and for the scale-based answers, please mark the points corresponding to your answer.

18. What is the area of publicly accessible spaces not subject to access control that are managed by the initiative, on average per year, in hectares

0	<0.1	0.1–0.5	0.5–1	>1
0 pts	1 pts	2 pts	3 pts	4 pts

19. Are the following elements of green infrastructure available within the garden/ initiative? ....._

    Rooftop garden?_ _ _
    Green walls/vertical gardens and landscaping?_ _ _
    Rainwater collection and irrigation system?_ _ _

20. Are the following shared social spaces and elements available within the garden/ initiative boundaries:

    Gazebos?_ _ _
    Dining tables?_ _ _
    Fireplace/bread oven?_ _ _
    Bicycle racks?_ _ _
    Biking lanes?_ _ _
    Playground for children?_ _ _

21. Was the initiative/garden established on top of:

    Abandoned/ neglected space?_ _ _
    A place where the soil was originally covered (by asphalt/other pavement, illegal dumping site, etc.)? _ _ _
    Previously built-up space/destroyed buildings?_ _ _

RESULT FOR THE SPATIAL INDICATORS
Total number of points:___

Environmental Aspects

For each Yes answer, please mark 1 point in the blank space, and for the scale-based answers, please mark the points corresponding to your answer

22. Do gardening activities put a special emphasis on environmentally sustainable agri-technical or animal husbandry practices (the principles of organic farming, permaculture, etc.)?_ _ _
23. If the produce is marketed, is it or part of it labelled as an organic/clean/ environmentally sustainable product?_ _ _
24. Are composting practices applied on a regular basis in the garden? _ _ _
25. What is the share of the garden/initiative area occupied by flowering and/or essential oil plants (flowers, fruit trees, lavender, etc.)

0%	<25%	25–50%	50–75%	>75%
0 pts	1 pts	2 pts	3 pts	4 pts

26. What is the share of the garden occupied by trees? (considering the width of their crowns?

0%	<10%	10–25%	25–50%	>50%
0 pts	**1 pts**	**2 pts**	**3 pts**	**4 pts**

27. Is there evidence the soil and irrigation water in use are harmless (based on laboratory sampling for contaminants or pathogens)?_ _ _
28. Does the garden/initiative have the following elements?

Bird-feeders?_ _ _
Insect hotel?_ _ _
Water elements (pond/fountain/creek)?_ _ _
Drip irrigation?_ _ _

RESULT FOR THE ENVIRONMENTAL INDICATORS
Total number of points:_ _ _

How to analyse your results? Please enter the points for each of the indicators in its relevant segment of the radar chart according to the preset intervals. You do not need to be too precise, the aim is to find out if your initiative has a balanced impact on the elements of the main aspects of QoL or some of them are dominant over other.

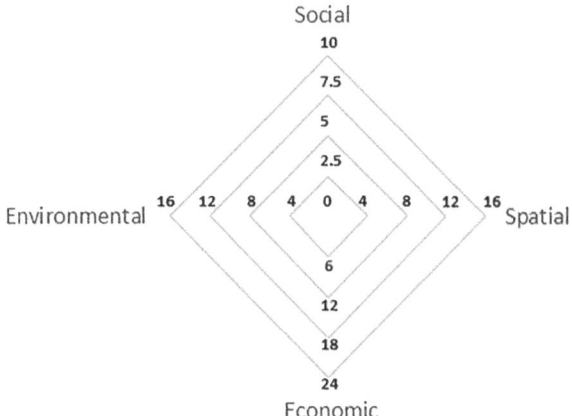

The manufacturer's authorised representative in the EU is Springer
Nature Customer Service Centre GmbH, Europaplatz 3, 69115 Heidelberg,
Germany. If you have any concerns regarding our products, please
contact ProductSafety@springernature.com

Printed and bound by CPI Group (UK) Ltd, Croydon, CR0 4YY
24/04/2026
02096308-0002